GREEN FACTS

THE GREENHOUSE EFFECT AND OTHER KEY ISSUES

GREEN FACTS

THE GREENHOUSE EFFECT AND OTHER KEY ISSUES

MICHAEL ALLABY

HAMLYN

First published 1986 in hardback
as *Ecology Facts.*
This revised softback edition published 1989 by
The Hamlyn Publishing Group Limited,
a Division of the Octopus Publishing Group,
Michelin House, 81 Fulham Road,
London SW3 6RB

ISBN 0 600 566 188

Phototypeset by Servis Filmsetting Limited, Manchester
Printed in Spain

Title page In the humid tropics the character of the forests changes as you climb from sea level into the high mountains. Above about 1000m there is almost perpetual cloud, and a belt of forest in which the canopy is fairly complete. Although there are some gaps, trees are often gnarled and twisted, and are shorter than at lower levels, and there is an abundance of liverworts, mosses, lichens, climbers, and epiphytes, some of which festoon the trees like cobwebs, This is 'cloud forest'. The photograph was taken in the Monteverde Cloud Forest Reserve, Costa Rica.

Two-colour illustrations and diagrams by Stephen Lings

Lito. A. Romero, S. A. – D. L. TF. 327 – 1986

Contents

INTRODUCTION

They say no one can check the advance of an idea whose time has come. It is as though some concept, some vision of the way the world is or might be, seizes the imagination of millions of people and changes the way they think, so that afterwards nothing can ever again be as it was before.

People have always cared about their surroundings. Our very countryside, that we like to think of as 'traditional' and 'permanent', was made by generations of humans who fashioned it partly to supply their needs but also to express their idea of the universe and their own place in it. You can see this more clearly in some parts of the world than in others, but only because the transformation takes time. The places where it is most evident are those that have been inhabited longest, such as the lowland farming areas of Europe and Asia.

We care, and when our surroundings become seriously degraded or polluted, we protest, often loudly. The English have been trying to conserve woodland at least since the twelfth century and the first smoke abatement law dates from 1273 – because of the belief that food cooked over burning coals caused illness and death, a belief that would have seemed eccentric twenty years ago but would receive much medical and scientific support today. The burning of coal was prohibited in London in 1306 because of fears of air pollution. As the manufacture of goods was industrialized during the eighteenth and nineteenth centuries environmental degradation accelerated and the protests grew louder.

Until very recently, however, environmental protection was subordinate to other interests, especially

The Cotswold countryside. The English landscape has been fashioned over several thousand years by generations of farmers, but modern farming has brought new and rapid change.

to those that were profitable. 'The state of the environment' did not inspire a popular movement until the 1960s, but once it had begun the movement grew so rapidly that within ten years most governments had been compelled to establish departments or ministries to deal with environmental issues, the United Nations had held a conference on the human environment, and a new United Nations agency had been formed, the United Nations Environment Programme. It was an idea whose time had come.

How to save the world

The environmental movement is political, in that it seeks to achieve its reforms mainly by means of legislation. This requires it to operate on two levels. It must persuade politicians of the need for the changes it advocates. It must also show that voters will continue to support those politicians who favour environmentalism.

At both levels the environmentalists must present their arguments in a sensational form. These must be clear, simple enough for anyone to grasp quickly, and the consequences of ignoring them must be made to seem dire. This is not meant as a criticism of environmentalists: it is simply a fact of political life. Unfortunately, it can lead to some confusion because the facts are seldom as simple as they are made to appear, and they are not always easy to interpret. Facts are sometimes disputed, and even where the facts are accepted alternative interpretations may lead to more than one conclusion. Nor are the warnings of dire consequences usually justified. The environment is much more robust than it may seem and it is unlikely that any damage to it will destroy it permanently. Several times in the past ice sheets have destroyed all plants and scoured all soil from any land whose latitude is higher than about 50°, yet when the ice retreats life returns.

Environmentalists wish to reform the relationship between human beings and their non-human environment. In a general sense, such relationships among living organisms and between them and their non-living environment are the subject matter of the scientific discipline called 'ecology'. Environmentalists make free use of the word 'ecology', and claim scientific support for all the statements they make, but although scientists may make emphatic, unequivocal statements when they are campaigning for causes they support, in private and in their own work they are much more cautious.

Scientists are people, and make moral judgements, but science itself is not concerned with morality or politics. To the ecologist one set of relationships among organisms may be different from another, but cannot possibly be 'better' or 'more natural'. 'Better', 'worse', 'natural', and 'unnatural' are words that have no scientific meaning. Environmentalists urge us to change the way we live, and often use the word 'ecology' as though it were the name given to a political or moral philosophy rather than a branch of science.

'Environmentalism' and 'ecology' are two words in common use, and there is a third, 'conservation'. The conservation movement is older than the environmental movement, and distinct from it. The two movements sometimes make common cause but it is possible to be a conservationist without being an environmentalist, or an environmentalist without being a conservationist.

Ways through the maze

In this book I have tried to chart what may look like a maze. I begin by explaining what the science of ecology is. The explanation is necessarily brief, but there are many excellent textbooks, the subject is taught in schools, and it is not difficult to find further information. Then I try to show how this branch of science is related to the popular environmental movement, how that movement is related to the conservation movement, and how both began and what they have achieved.

No book can hope to deal with all the issues that concern conservationists and environmentalists. There are too many of them and they change too rapidly. I have tried instead to illustrate the kinds of issues that concern people, and the way they are approached. To do that I have selected a list of nine broad subjects, starting with industrial pollution and ending with the world food and population problems.

Each of the subjects illustrates a particular point. In some cases the apparent problem may be unreal. Or the central issue may have been misunderstood, so that the target of environmental protest is actually the wrong one. In other cases the problem

Early cultivation for the spring sowing of barley on a large modern field near Marlborough, Wiltshire.

may be real, but so complex and with so little known about it that we have no choice but to wait until we have more information before we decide what should be done. In some cases real problems have been identified and solutions are in sight. They are success stories that should encourage us to tackle others, issues that only now are being recognized but which will concern us much more in the next few years.

What can we do?

The achievement of any reform requires action on the part of those whose behaviour must change, in some cases with legislation to ensure fairness or to compel the more wayward to adopt the standards accepted by the majority. Once achieved, reforms can be sustained only through vigilance. There is little any one of us can contribute as an isolated individual, but much we can do if we are prepared to work together.

If you feel concern about the condition of the environment or the conservation of landscapes or wildlife generally, or if there is a particular local abuse you wish to see remedied, then my first advice is to join a local branch of a national conservation organization. Among its own members the conservation movement has a vast store of specialist information, and what its members do not know in most cases they can discover through their contacts with professional scientists. The conservation movement is well respected, and it is respected because it is reliable in matters of scientific fact and interpretation.

Then you may wish to join an environmentalist group to campaign for a particular reform. The group will be essentially political and its point of view will be partisan, but you will be able to draw upon considerable campaigning experience, and you may need that if you seek even a local reform.

There is a final warning. At all times you should avoid allowing others to do your thinking for you – however worthy they may seem. Scepticism is a formidable tool, provided you apply it impartially to the opinions of everyone. This means you owe it to yourself to hear both sides of each argument, and to arm yourself with facts rather than opinions, which may involve some study. The 'expert' and the 'official spokesman' also deserve a hearing. They may know what they are talking about – they could be right.

CHAPTER 1

What is 'ecology'?

In a book published in Berlin in 1866, the German biologist Ernst Haeckel (1834–1919) wrote that each individual living organism is the product of co-operation between its environment and the body it has inherited. He described this co-operation as 'oecology', from the Greek word *oikos*, meaning 'house'. Aristotle wrote about the relationship between organisms and their environment and many people studied it over the centuries, but it was Haeckel who coined the name by which it is known to this day. Ecology (to give it its modern spelling) is the study of the relationship between organisms and their environment.

Like most biologists of his day, Haeckel was deeply interested in evolution and he became an enthusiastic Darwinist, for Darwin's theory of evolution by natural selection is an ecological theory. It holds that minor differences among individual members of a particular population of a particular species equip some better than others to thrive and reproduce in the environmental circumstances in which they live. The inherited differences provide the basis for natural selection, and the selection itself results from pressures originating in the environment. Despite this obvious connection, Haeckel's 'oecology' was largely overlooked and remained forgotten for years. It was not until the turn of the century that biologists, at first mainly in America, began to study the 'co-operation' seriously. The word 'ecology' dates from 1866, but the science of ecology did not begin until around 1900.

Science is concerned with the observation of the world around us and with devising plausible explanations for the events that are observed. There is

A fine example of English, or pedunculate oak (Quercus robur) *in its autumn colours. A mature tree like this provides food and shelter for so many other species it is an ecosystem in itself.*

nothing very mysterious about it, but its explanations have to be real explanations, and really convincing, and they are rarely final because each explanation raises new questions.

It is not until people working in a particular branch of science have accumulated large numbers of observations and have devised satisfactory explanations for them that they can start to make predictions. This takes time. Ecologists are still accumulating observations and devising explanations. Their science is growing rapidly, but it does not yet allow them to make many precise, reliable predictions. It is not their fault, or that of their science, but only that their discipline is very young.

This explains some of the worry and confusion that surrounds environmental issues. We are aware that problems exist, but uncertain about how serious they are or how to solve them without creating still more problems with our solutions. We need more information, and when scientists respond to environmental controversies by demanding more research they are not usually trying to evade the issue, or enhance their own careers. They really do need to know more. What ecologists have learned so far can be summed up very simply: the world is a great deal more complicated than anyone thought.

Communities and ecosystems

Ecologists study the relationships between organisms and their environment – but what is their 'environment'? To a large extent it consists of other organisms. So ecologists study whole communities. What, then, is a 'community'? You might study all the plants and animals in your own garden, for example. If the hedge surrounding the garden on three sides, and the house on the fourth, form a boundary to distinguish what is inside from what is outside you might suppose that your garden is a distinct community. It is not really sealed, however. Insects enter and leave freely. Birds fly in and out, and some even nest in the garden, if they manage to escape the cat from next door which hunts them. Are the insects, birds and cat members of the garden community?

Your own garden is a very rewarding place to study ecological relationships, but it is not the place to study a complete community because the hedges and house are not enough to separate it from the

Little owls (Athene noctua) *looking out from their nest hole in the base of a great oak tree. The owls are three weeks old.*

almost identical gardens nearby. If you are to study a particular community you will need to define it much more precisely.

Ecologists use the word 'ecosystem' to help them. An ecosystem is an area you can define, because in some important way it really is different from the area around it.

The use of the word 'system' is significant. To a scientist, a system is a discrete assemblage of parts which are related to one another in such a way as to maintain the integrity of the whole. You might think of your own body as a system, in which each organ, each muscle and blood vessel plays its part so the body continues to function as a whole and the whole can be regarded as an entity in its own right. Your body is much more than the mere sum of its parts, and so is an ecosystem.

Systems are self-regulating by means of 'feedback' relationships. In an ecosystem containing grass, rabbits and foxes, for example, if the rabbits increase in number there can be more foxes to hunt them, so fox numbers will also increase and rabbit numbers will be checked. While they are more numerous, however, the rabbits will also eat more grass. This may reduce the amount of grass, so limiting further increase in the number of rabbits because there is insufficient food for more of them. These are feedback mechanisms, in which an increase at one point in the system causes a reaction elsewhere in the system which 'feeds back' to the original point to restore it to its former value.

An ecosystem need not be large but it can be. A forest might be studied as an ecosystem, because it has a definite edge, and the land surrounding it is not forest, so it is different. That is a large ecosystem. A single tree may also be regarded as an ecosystem. You can see it clearly, and it has a boundary, an edge on one side of which there is tree and on the other side of which there is air or soil. The tree itself is the place where birds nest, insects feed, mosses and lichens grow. The tree is alive, but it is also a place in which a community of plants and animals lives. A pond, a puddle, even a single drop of water held on a leaf can be studied as an ecosystem.

An ecosystem is not truly closed. Sunlight and rain enter it, surplus water drains from it, plant nutrients, moving through the soil, may enter or leave it, and the living organisms themselves are not so confined as they may seem. Seeds can move into the area or from it, carried on the wind or by animals, and animals, especially those that are airborne like insects, birds or bats, come and go.

Ecotones and biomes

At their edges many ecosystems shade gradually into the area adjacent to them: the boundary is not abrupt. A woodland plantation may consist of trees growing right up to the fence and no trees at all in the field beyond, but in woodland that has developed without human interference the trees become more widely separated and grass and other herbs grow among them, and the grassland penetrates in long 'fingers' among the trees. Such a boundary area is called an 'ecotone' and because it supports plants and animals typical of the ecosystems to either side it is especially rich in species.

Natural woodland exists in three dimensions, its plants forming distinct layers. This drawing shows four, but there can be as many as seven. Different species of birds nest in different layers, but about half of all woodland birds feed on the ground.

*Left The beech and oak high forest of Ebernoe Common, Sussex, seen early in autumn. Contrast this oceanic-temperate ecosystem with that of the semi-arid continental Sonora Desert of Arizona and New Mexico (**above**). The character of most ecosystems is dictated by the climate.*

A forest may be regarded as an ecosystem, but in some parts of the world the same mixture of trees, shrubs and herbs extends over vast areas. Such an area of forest stretches across Canada, northern Europe, and northern Asia. Farther south there are similarly vast regions of grasslands – the North American prairies and the Eurasian steppes. South of them there are deserts, and in the tropics there is another region of forest, occurring in America, Africa, Asia, and the northern tip of Australia. These great ecological regions of the world are called 'biomes'. The entire living component of the planet, covering almost the whole of its surface and atmosphere, is called the 'biosphere'.

Relationships and change

The ecologist is interested in relationships, and in any ecosystem there are many. The most obvious relationship is between an organism that is eaten and the organism which eats it, but that is only a start. There are relationships between males and females, between adults and young, between dominant and subordinate individuals, or between resident animals and 'visitors' from outside. There are opportunist plants and animals that appear when there is food and move somewhere else when it has been eaten and so form temporary relationships with more permanent residents. There are relationships that include animals which protect themselves by looking or behaving like other, quite unrelated animals that are dangerous or inedible. There are several kinds of commensal relationships in which individuals of different species live closely together, and there are parasitic relationships. These occur at several levels, because some parasites, called 'hyperparasites' parasitize other parasites.

As though this were not complicated enough, ecosystems do not remain the same. Except near the equator where the weather stays the same throughout the year they change with the seasons, and the more marked the seasons the more marked are the changes. Nor do they appear suddenly, fully formed. In most of Britain, for example, if you leave an area of countryside alone shrubs and a few small trees will appear in it after a year or two. As they grow, the grass around them will die back because grass can tolerate many things, but not deep shade. The trees will grow taller and provide shelter from the wind for more trees. Eventually, the area, which may have started as a field of grass, will become woodland. Then the composition of the woodland will change. New tree species will arrive, and others will fail to establish seedlings so that when the mature trees die they are not replaced. It will take many years, but a succession of communities will follow one another until at last the woodland stabilizes. This 'climax' community will endure for a long time, perhaps for centuries, but in the end it, too, will disappear and be replaced by a quite different community if for no other reason than that the climate itself changes. We in Britain are living in what most climatologists believe to be an 'interglacial', a warm interval between two glaciations (ice ages). One day, thousands of years from now, the ice will return. The development of communities over long periods, through successional phases such as these, is of great interest to ecologists.

Bringing order out of confusion

The subject matter is bewildering in its complexity, but some order must be found in it if anyone is to study it. Relationships based on food are as good a place to start as any.

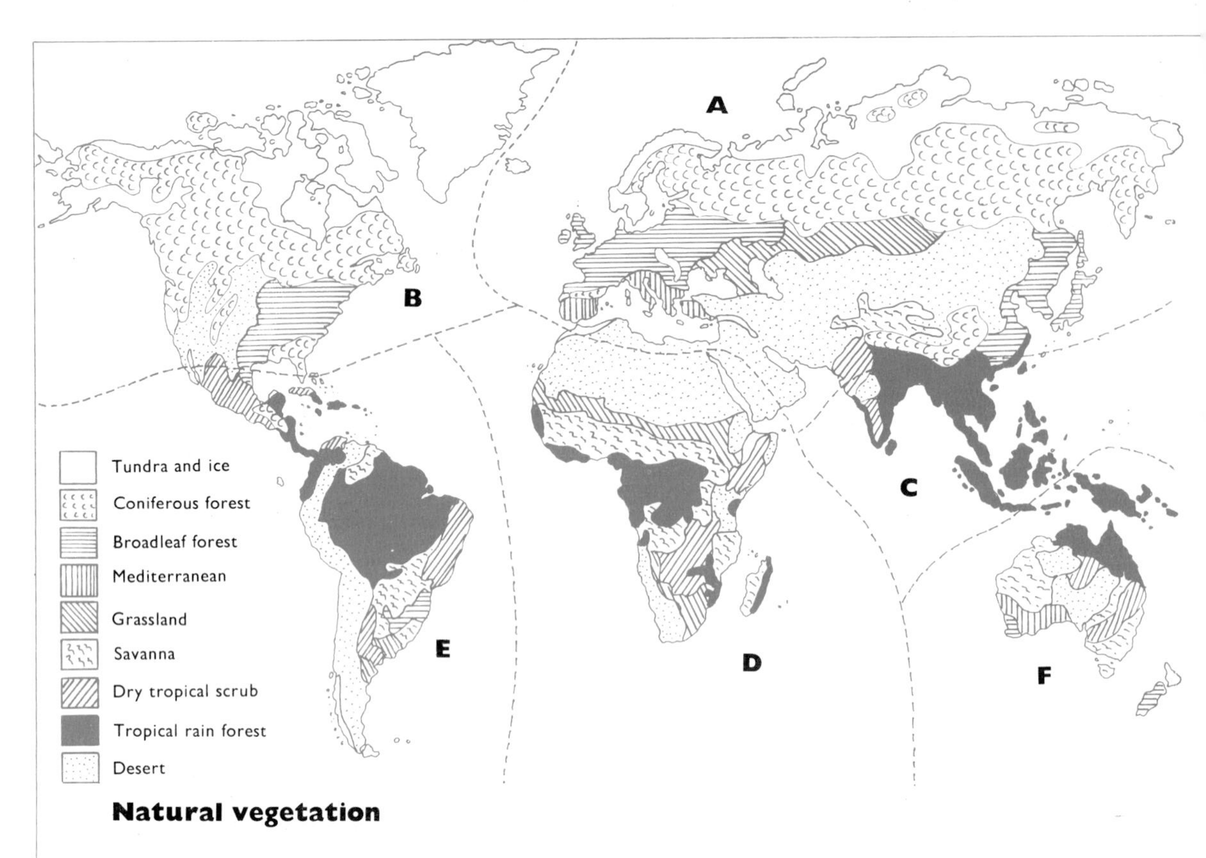

Natural vegetation

Zoographical regions

The natural distribution of animals is such that quite distinct groups occur in particular parts of the world. This observation has led to the division of the world into three zoographical regions, called Arctogea, Neogea, and Notogea. Arctogea is further subdivided into four smaller regions, the Palaearctic, Nearctic, Oriental, and Ethiopian.

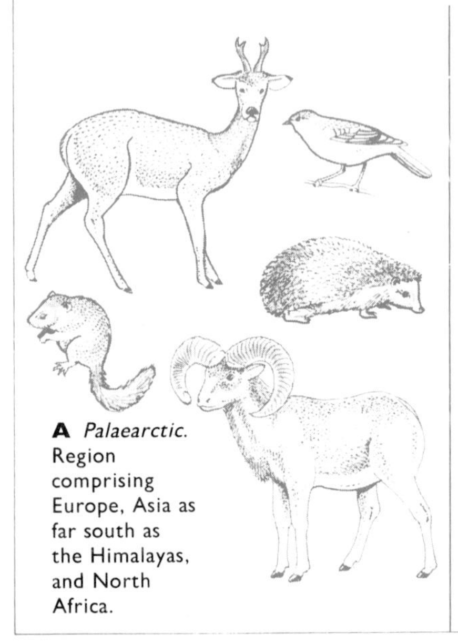

A *Palaearctic.* Region comprising Europe, Asia as far south as the Himalayas, and North Africa.

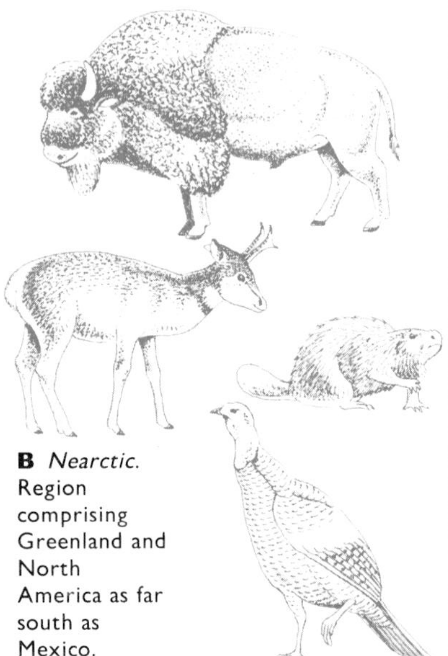

B *Nearctic.* Region comprising Greenland and North America as far south as Mexico.

C *Oriental.* Region comprising India, Indochina, Malaysia, the Philippines, and those Indonesian islands lying to the west of Wallace's Line (which separates the Oriental and Notogean regions).

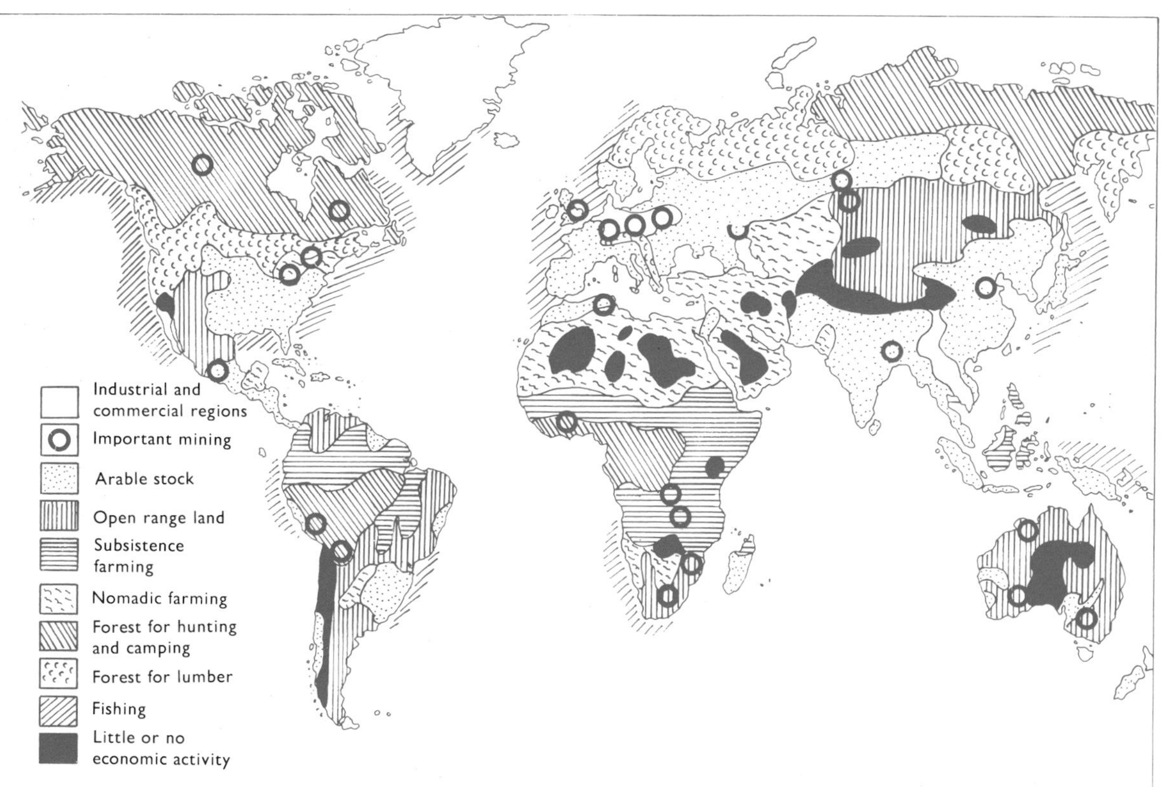

Chief areas of productive and non-productive land

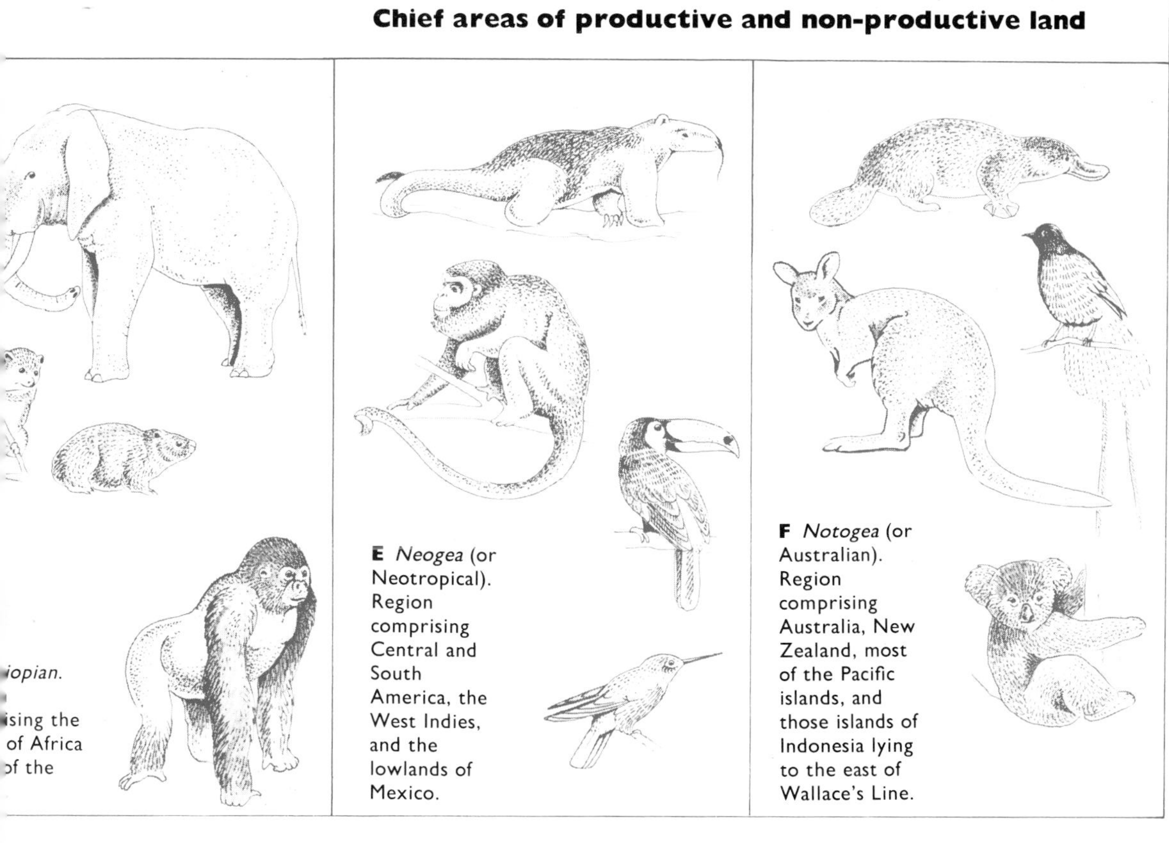

iopian.

ising the
of Africa
of the

E *Neogea* (or Neotropical). Region comprising Central and South America, the West Indies, and the lowlands of Mexico.

F *Notogea* (or Australian). Region comprising Australia, New Zealand, most of the Pacific islands, and those islands of Indonesia lying to the east of Wallace's Line.

In any ecosystem there are certain organisms that manufacture the substances from which they are made, using as raw materials simple chemical compounds they obtain from the air and from the water in the soil, or even from rocks. Many micro-organisms do this, but the most familiar examples, and ones we see all around us every day, are green plants. With the help of chlorophyll and the energy of sunlight they break water into hydrogen and oxygen, release some of the oxygen, and combine the hydrogen with carbon dioxide to make sugars. The process is called photosynthesis. They make proteins and other necessary substances from the molecules of chemical salts they absorb in solution through their roots. Green plants and those micro-organisms which either photosynthesize – and some of them use versions of chlorophyll that are not green – or obtain their energy by other chemical processes such as fermentation, are converting simple, inorganic substances into complex, organic substances. They are called 'autotrophs' (self-nourishers).

An organism that is not an autotroph is called a 'heterotroph'. The name means 'different nutrition' and a heterotroph must obtain substances to supply it with energy – fuel – and to make and repair the cells of its own body in the form of those complex organic compounds. The only way to obtain those is by eating other organisms.

Already we can divide all the organisms in any ecosystem into two groups: autotrophs and heterotrophs. Heterotrophs eat autotrophs, but some of them eat one another; among them there are herbivores and carnivores, not to mention omnivores like ourselves, which eat both autotrophs and other heterotrophs. Now we have several groups and we can begin to examine the relationships among them in two ways, as groups or as individuals.

Food chains, webs and pyramids

Think of them as individuals and it is possible to draw a diagram of what eats what. At its simplest level you might imagine a rose, which produces hips in the autumn. The hips are eaten by mice, and nearby there is an owl which feeds on mice. That is a 'food chain', but in the real world 'chains' turn rapidly into very involved 'food webs'. In the first place there are not just roses, but many plants, and they all

Top *A caterpillar of the pale tussock moth* (Dasychira pudibunda), *resting on an acorn in a pedunculate oak tree.* ***Above*** *The spotted flycatcher* (Muscicapa striata), *a bird of parks and woodland edges, hunts flying insects.* ***Right*** *Red foxes are among the most versatile of predators. This is the North American* Vulpes fulva, *and it has just killed a cottontail rabbit* (Sylvilagus).

The kind of food web that exists on open heathland with scattered trees. Insects feed on plant material, and are eaten by predators such as centipedes, spiders, birds, and small mammals. Small mammals eat seeds and insects, and are eaten by predatory mammals and birds of prey, such as the short-eared owl. The owl will also eat birds smaller than itself.

produce seeds. Insects feed on the plants, and invertebrate predators, such as spiders and centipedes, hunt the insects. Shrews and voles eat seeds as do the mice, and mice and shrews also eat some insects. The owl eats some insects as well as mice, voles and shrews, and there is a snake not far away which hunts these small mammals. The animals which eat invertebrates eat the predatory ones as well as the herbivores.

Ecologists study such food webs because they are interesting, and may be important. It might be, for example, that people were worried about the decline in the population of owls. Owls eat small mammals, so the decline might be due to a shortage of mice, voles and shrews. That shortage might be due to a shortage of seeds or other plant material, which in turn might be due to the careless use of a herbicide nearby which had killed some of the plants, or to a lack or surplus of water caused by farm drainage. The disappearance of some of the owls might be the first sign people had of a change in a local beauty spot, but the ultimate cause of that change might have nothing to do directly with the owls themselves or with their prey.

If you decide to study the living organisms as groups rather than individuals you will need another technical term, 'biomass'. This is simply the total weight of all the organisms in an ecosystem or in some part of it. You might work out the biomass of all the autotrophs, then of all the herbivores, then of the carnivores and omnivores. This kind of calculation is used to determine how much new organic material an area produces in a season, its 'productivity'. A decline in productivity may be an early sign of trouble or of major change.

The autotrophs can be regarded as 'producers' and heterotrophs as 'consumers', and the consumers occupy several 'levels', as primary consumers (herbivores), secondary consumers (omnivores and carnivores), and tertiary consumers (carnivores that prey mainly on other carnivores). The levels are called 'trophic levels', and if the biomass at each trophic level is represented by a bar of a certain length the bars can be arranged one above the other, like a bar chart set on its side but with each bar centred, so that it forms an ecological pyramid. It may represent biomass, or food available at one level for the level above, or the food may be converted into energy. All such pyramids have roughly the same shape: each level is much wider than the level above

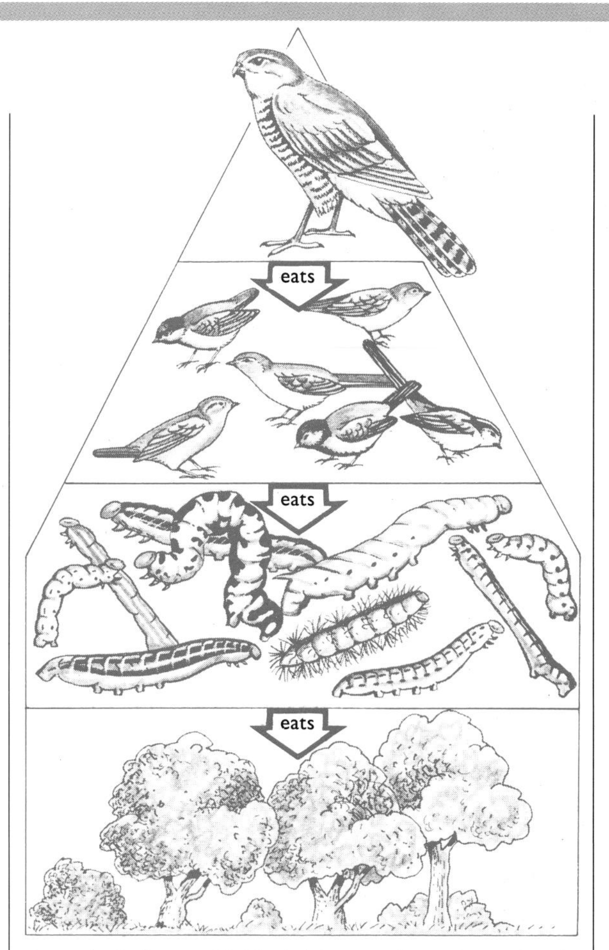

An ecological pyramid. The total mass of organisms at each level is roughly one-tenth of that at the level below.

it. In energy terms, for example, the efficiency varies widely from one kind of ecosystem to another, but on average no more than about 0.1% (one thousandth) of the solar radiation reaching the ground surface is absorbed and used by green plants. About 0.015% of the total radiation is used by herbivores, 0.0003% by carnivores, and 0.00004% (four millionths) by 'top' carnivores.

This is a very important ecological observation. It means that a disturbance at one level will have a large effect on all the levels above it. This is an example of positive feedback, in which a change in one part of a system has effects in another that feed back to amplify the original change. Positive feedback leads to instability and sooner or later to deterioration of the system itself. It explains why it is so rare for carnivores to specialize in hunting other carnivores: it is not because they are difficult to hunt, but because there are so few of them. There is not enough food, and there cannot be because of the inefficiency with which energy is passed from level to

level. It also explains why there are so few very large carnivores in the world: they need a great deal of food to sustain them. This means they must hunt in a large area. If the carnivores themselves were larger the area supplying their food might be too large for them to patrol.

Counting, weighing and measuring

Ecologists are more concerned than most biologists with counting, weighing, measuring, and producing numbers and they rely very heavily on statistical techniques, these days often applied with the help of computers. Computers allow ecologists to represent mathematically many of the relationships within an ecosystem, feed those relationships into a program as equations, add in the values obtained from observation and run the resulting 'model' to simulate ways in which that ecosystem might respond to change. The answer is theoretical, but it may provide useful guidance when large-scale developments are proposed.

How do you work out the number of species in an area, and the number of individuals of each species? You could go through a wood and count every tree, every shrub and every blade of grass, but it would take a long time and, while the plants do not move about, the animals do. You could not count them in this way. You need ways to sample populations. There are several you might try in your own garden, or in some area of countryside to which you are allowed access.

To discover the commonest plant species you might make a number of transects. First prepare a map of the area you plan to survey, then draw straight lines across it at intervals. Mark those lines on the ground with tape or string, and list all the plants that touch the line. You will compile lists of names and numbers, one list for each line, and by performing various mathematical manipulations you will be able to calculate the relative frequency of each of the plants you found.

You might complement this by marking points on your map as randomly as you can manage. Find each point on the ground and place over it a square frame. The usual size is a one-metre square. Then list everything inside the frame. This is called quadrat sampling. You can make it simpler by using a frame that surrounds a grid made of wires, and noting everything you find at each of the intersections in the grid. Again, mathematical manipulation of the information you gather will lead you to a fairly reliable estimate of the population of the whole area.

An ecologist, in Epping Forest, recording vegetation with a point quadrat, noting each plant, or bare ground, touched by pins lowered through the frame.

Female large blue butterfly (Maculinea arion) *on a wild thyme plant. The large blue was the first butterfly to become extinct in Britain for a century.*

Plants are quite easy to count. Animals are more difficult because they are highly mobile. Usually animal populations are calculated on the basis of trapping techniques. You can improvise the equipment you need to sample the animals living in the soil, although you will need at least a powerful lens and preferably a microscope to be able to count the smaller ones, and with a little ingenuity you might improvise equipment for sampling insects. You should not try to trap mammals. Ecologists do this, but they use specially designed traps that will not injure the animals that enter them, and they have to be preset for the animals they are to catch. Shrews especially will starve to death in a few hours if they cannot forage for the food they need.

It is great fun, and very instructive, to find out just how many wild plants and animals there are living even in such an apparently tame area as a town garden, but you will soon discover that it is one thing to 'count heads' and quite another to work out what the numbers really mean. If you are to practise ecology, even as an amateur, you will need a good grasp of the relevant mathematics. Once you acquire that, however, you can put the home computer to good use.

A census of species allows ecologists to identify those which are rare. In the past many species have become extinct because no one knew they were there at all, or that they were of value. Most extinctions occur not because humans set out deliberately to destroy particular species, but because their habitat – their environment – undergoes a radical change. At one time it might have been possible to argue that a rare species can be kept alive by itself. Perhaps you could put a fence around it, build houses and factories over the rest of the surrounding area, and the species would live on quite happily. Ecologists have shown why this rarely works in the wild. You can keep animals in zoos and plants in botanical gardens, but they require constant expert attention. In the world outside species interact with their environments through networks of relationships so intimate that if they are to survive their environments must also survive.

The disappearance of the large blue butterfly

Changes that prove damaging in the end may seem unimportant at first. In September 1979 the large blue butterfly (*Maculinea arion*) was declared extinct in Britain. It used to live on the chalk downs and its life cycle was complex. The newly hatched larvae looked like the flowers of wild thyme, the plant on which they fed. As they grew they moulted their skins and after the third moult they left the thyme and waited in the grass for ants of a particular species. When stroked by the antennae of an ant the caterpillar exuded a sweet substance on which the ants fed, and the ants would carry the caterpillar back to their nest where, in return for its sweet exudate, they would shelter it and feed it – on their own larvae. The caterpillar would hibernate through the winter, pupate in spring, and crawl to the mouth of the ant nest in midsummer, to emerge as an adult.

The key to its life cycle was the wild thyme. This could not grow if it was shaded by surrounding grass, but the grasslands were grazed intensively by sheep and rabbits. Changes in farming reduced the numbers of sheep, but this allowed the rabbit population to increase. It was myxomatosis that depleted the rabbit population. As the rabbits disappeared the grass grew taller, shaded the thyme, and so the thyme disappeared and the large blue butterfly with it. The butterflies, you might say, fell victim to myxomatosis in a roundabout kind of way.

Ecological advisers

Although that disappearance was not foreseen ecologists spend much of their time identifying habitats that are important for the species they support.

Other problems are also predicted and these days ecologists are usually consulted when any major change in land use is contemplated. If those favouring the change do not take ecological advice, their opponents will.

Ecologists can issue warnings, and although their predictions are usually imprecise they are much better than no predictions at all. They can calculate the size and reproductive capability of a population of animals that are being hunted – such as whales – to determine the size of the kill that can be maintained year after year without depleting the population. They can warn of environmental changes that may reduce populations of animals that are popular or of economic value. They can also help diagnose problems and this makes them deeply involved in many environmental arguments, such as those surrounding the issue of acid rain.

The role of the ecologist in planning is not wholly negative, however. Ecologists can also advise on ways in which areas may be improved to allow them to support a richer wildlife and this forms a large part of their work.

The extractive and manufacturing industries move into areas and after a number of years they leave to go elsewhere. Gravel pits and quarries are worked out and abandoned, spoil heaps and waste tips are built and then left, factories close and are demolished, and in the past local residents have been left with unsightly, derelict and sometimes dangerous land for which there was no obvious use. The soil might be poisoned, tips might catch fire or leak substances that could poison rivers, pits and quarries would fill with water. Today such areas as these can be reclaimed, usually to become amenity areas. They can be landscaped. Soil can be imported or made. Trees and other plants can be sown. Gravel pits can be allowed to fill with water, to become artificial lakes for fishing or boating.

Workers clearing a stretch of the Basingstoke Canal in Hampshire. Most British canals fell into disuse and became clogged and polluted when freight moved to rail and road. The Basingstoke, built around 1810, is being transformed into an attractive amenity.

Their reclamation is mainly a matter of landscape design and engineering, but ecologists can and do help. They can pick out areas in which particular communities of plants might be encouraged to develop, attracting particular species of animals. They can suggest ways in which small areas of 'wilderness', where there will be little or no human interference, can be linked to form corridors and networks that allow animals to move about without leaving the shelter and food supply they need, and yet without detracting in the least from the amenity value of the area as a whole. Enhanced in this way, a park that is made on the site of old factories, right in the centre of an industrial city, can also be a classroom in which students can observe the natural world from which their urban environment otherwise excludes them.

A scientific jack-of-all-trades

Ecology is a branch of biology concerned with relationships among species and between species and the physical and chemical environments they inhabit. The ecologist must be part biologist, part chemist, part physicist, part geologist, and an accomplished mathematician. Although much of their work appears to be abstract and their science is still too young and provides them with too little detailed information to allow them to make confident predictions, already ecologists are in demand to advise on the wise use of our natural resources, and most especially our landscapes, and to warn us of human activities that might inflict serious damage on non-human communities.

CHAPTER 2

Human ecology

Ecology is the study of the relationships among living organisms and between those organisms and their inanimate environment. If the study is narrowed to humans, human modification of the environment, and the effects of such modifications on humans themselves, the result is 'human ecology'. It is hardly a scientific discipline because it is not really legitimate to isolate one species in this way when considering a total system, but it does permit the environmental effects of purely human activities to be measured. Then, if they appear to be adverse to humans themselves, correctives can be proposed.

It is anthropocentric – 'human-centred' – and 'environmentalist'; the devising and advocacy of reforms intended to improve the quality of the environment is similarly anthropocentric. This is not to condemn it. We might campaign, for example, to make certain factories clean their effluents before discharging them into a river in order to protect the health of people living nearby and to change the river from something little better than an open sewer into an attractive amenity. People are the intended beneficiaries, but if the campaign succeeds and the river becomes cleaner plants and animals will be able to colonize it, so they will also benefit.

We need to be clear about our motives because if we deceive ourselves we may be trapped into making a quite false distinction between humans and non-humans – the very distinction some environmentalists claim to oppose. They assert that humans are part of the ecosystems they inhabit and are entitled to no special privileges. Non-humans have equal rights, at least to stay alive. Yet the main threat to the non-humans is perceived as coming from

Smoke, exhaust gases and water droplets combine to obscure the dawn light at a Teeside chemical plant and refinery. A few decades ago most industrial areas looked like this.

humans, and so the very division that was denied is assumed to exist, but in a new form. Instead of being uniquely blessed, humans are regarded as uniquely wicked, and it is but a short step further to the assertion that humans and only humans alter the world to suit themselves. Humans are uniquely selfish. If this is true, how can we allow ourselves to do anything at all in the world? It seems like an argument for mass suicide – were it not for the pollution that would cause!

Humans and non-humans

The truth is that every living organism alters its environment. Indeed, the modification of the environment is one way by which we may define the word 'living'. If you were investigating another planet to discover whether it supported life you would not travel the surface looking for living organisms directly unless they were so large and obvious as to be unmistakable. You would examine the air, soil and water to see whether their composition departed from what the laws of chemistry and physics predict. You would look for perturbations that could be explained only by assuming the presence of living organisms. You would look for environmental modifications.

On our own planet such modification occurs on a vast scale. Our atmosphere contains oxygen as a by-product of photosynthesis – it is released by green plants. The great grasslands of North America, Africa and central Asia may well have been formed by animals. Repeated fires, possibly caused by humans as a way of driving game, or by accident, may have destroyed trees. Grasses cannot tolerate deep shade, but the removal of trees allows them to flourish. The proliferation of grasses would increase the food supply for large grazing herbivores such as bison, cattle, deer and antelopes, and as they grazed they would have destroyed tree seedlings and so prevented the regeneration of the original forest.

A road cut through the tropical forest in western Brazil by simply clearing away the vegetation, has been partly destroyed by erosion as fierce rains have beaten down on the unshielded soil and swept it away, into the nearest river.

Mount St Helens, Washington, erupted explosively on 18 May 1980, releasing ash, stony fragments and pumice, and causing landslides, flooding and mud flows. The ash plume circled the Earth, but had no significant climatic effect.

There are countless similar examples and they provide much of the basis for ecological studies. An ecosystem, after all, is the organisms which comprise and have made it.

Humans are animals and like all living organisms we, too, alter our environment to suit ourselves. Those who regard all human interference with the environment as somehow 'unnatural' are profoundly mistaken unless they really do regard humans as in some sense outside or beyond the reach of 'nature' (whatever that means). Alternatively, some people consider that at some time in the past humans lived 'naturally', without interfering with their environment, but that today we live 'unnaturally'. This view is no less mistaken than the first. Humans have always altered their environment, often on a large scale and in ways that today we would consider detrimental. No great technological expertise is required to clear a forest and so cause severe soil erosion, or to create a desert by farming badly, or even to obtain meat by driving entire herds of animals over cliffs and leaving the surplus corpses to rot.

As a final position, may we defend the proposition that it is only humans who pollute the environment? Pollution is the introduction into air, water or soil of substances that are harmful to some or all living organisms. Alas, even this position cannot be held. Volcanoes eject vast amounts of chemicals into the atmosphere, some of them identical to those emitted by our own factories but in much larger amounts. We manufacture many powerful poisons, but poisons no less powerful exist independently of humans. Even plutonium, the radioactive metal produced in nuclear reactors and released by humans into the environment, occurs naturally and, millions of years ago, in Gabon, uranium-bearing rocks were brought together in such a way as to form a critical mass of uranium. A crude thermal nuclear reactor generated heat and produced plutonium. It caused no serious contamination, no harm to any organisms, but it happened.

The greatest pollution incident in the history of the planet had nothing to do with humans at all. It happened about 1.5 billion years ago, when free oxygen began to accumulate in an atmosphere that previously had contained almost no free oxygen, and that accumulation itself was the result of contemporary environmental management by living organisms. Atmospheric carbon dioxide was absorbed by photosynthesizing plants. It was used by the plants to make sugars and oxygen was released as a by-product. When the plants died and decomposed the carbon should have been oxidized back into carbon dioxide, so removing the gaseous oxygen, but instead carbon was buried, in ways that are explained in more detail in chapter 5. One consequence was that the climate of the world was held constant. The other was that gaseous oxygen accumulated in the air.

Oxygen was, and still is, intensely poisonous. It reacts with many substances and inside a living cell it can wreak havoc. Probably most of the organisms alive at the time were destroyed. Others took refuge in environments from which oxygen is excluded, such as muds and, later, the guts of large animals. Others evolved ways to use oxygen as a source of energy, but even for them surplus oxygen had to be rendered harmless and removed. Much of the biochemistry of our own body cells is devoted to the

Above The Merced River and Yosemite Valley in the Yosemite National Park. The Park occupies more than 300,000 hectares, of mountains, forests and alpine tundra.

Right Haydan Valley, Yellowstone National Park. The Park occupies 900,000 hectares and contains abundant wildlife and the world's largest geyser area.

detoxification of oxygen, and it is only partly successful. The metabolization of oxygen releases very small amounts of hydroxyl radicals (OH), hydrogen peroxide (H_2O_2) and other substances which are also produced by exposure to ionizing radiation. They cause cancer and they are what makes such radiation harmful. Over a lifetime an air-breathing animal is exposed to almost enough oxygen to cause cancer. This fact can be used to measure the hazards associated with particular levels of radiation exposure, and it also places an absolute upper limit on the lifespan of any animal since the lifespan of all animals is related directly to the rate at which their bodies use oxygen.

The beginning of conservation

Humans are different from all other species in one respect. We, and we alone, are able to think in abstract concepts and therefore to consider all the consequences of our activities. All living organisms modify their environments, but only humans can form societies and clubs to regulate that modification.

People have worried about the environment since classical times, and possibly for much longer. Roman writers complained of the soil depletion and erosion caused by 'modern' farming, and ancient Rome suffered seriously from pollution – and from traffic. The roots of our present conservation and environmental movements can probably be traced to the intellectual expansion and ferment known as the Enlightenment, in the eighteenth century, and they have taken two forms. One has been concerned with the study of the environment and has led to modern science on one hand and on the other to the amateur natural history movement and, through that, to the conservation movement. The second strand observed the condition in which humans were living.

That led in one direction to movements for political and social reform and in another to movements for the improvement of urban conditions.

The study of natural history also led to the romanticism which caused people to see the world as more than merely a resource for human use. For most of history 'wilderness' was a term of disapproval. Today people who live in very protected environments seek to preserve wilderness areas as places to visit. They have a new cultural and aesthetic value and the organizations formed to protect them are a distinct branch of the broader conservation movement.

The wilderness movement can trace its origins in Britain to the time of the poet William Wordsworth (1770–1850) and his friends. In America it began with the work of the Scottish-born naturalist John Muir (1838–1914), whose family moved to Wisconsin when he was 11, and Henry David Thoreau (1817–62). Muir founded the Sierra Club, devoted to preserving wilderness areas and to conservation generally. More recently it has become involved in environmental issues. Friends of the Earth, the environmentalist organization, is an offshoot of the Sierra Club. Even so, in the early days attitudes were ambivalent. When Thoreau lectured on the value of wilderness he was living in the civilized comfort of Concord, Massachusetts, and although he spent much time in the open air he had very little personal experience of living in true wilderness. When he did encounter it he was somewhat disenchanted. Walden Pond, which he immortalized, was close to Concord. It was not a wilderness area.

National parks, the first of which were being established at about that time – Fontainebleau, France, and Yosemite, California, both in 1864 – were not intended to protect wilderness for its own sake. They were seen as reservoirs either of natural resources to be held in trust for later economic exploitation, or as amenities, for public enjoyment. They were quite definitely intended for the use of people.

Sand dunes in a national park established in the Thar Desert, Rajasthan, India. Despite the extreme aridity, the desert supports many organisms adapted to its harsh conditions, and geologists study the natural formation and development of sand dunes.

The beginning of environmentalism

In 1962 Rachel Carson published *Silent Spring*, a book in which she criticized the use of pesticides in North America, arguing that this particular human activity was having a measurable harmful effect on non-humans, and especially on birds. The 'silent' spring, which she described on page 2 as a 'fable for tomorrow', was the one in which no bird sang, but many plants died and the fish in the rivers vanished. The study and protection of wild birds, flowers and fish was a matter for conservationists, but the reform of human activities was economic and political, and Carson warned that the consequence of the 'silence' of which she warned would be the failure of some food supplies – for humans. The issues had come together, the book provoked extraordinarily emotional reactions within the chemical industry, even before it was published and from people who at best had read only extracts. They added to the publicity surrounding it, which led in turn to the coalescence of a popular anti-pesticide movement. Environmentalism had been born.

The environmental movement was not advocating conservation, but it was concerned about the consequences of human activities on non-humans as well as humans. It had an interest in, and used some of the terminology of, the science of ecology. So sometimes it called itself the 'ecological' movement, but clearly such a term can have no scientific meaning. In this sense ecology, or human ecology, was more a political philosophy, and that is how the movement used it. People were exhorted to live 'ecologically' (scientifically there is no other way to live), and products which were held to cause the least harm to non-humans were described as 'ecological', or 'ecologically sound'. It became necessary to distinguish between the two uses of the word 'ecology'. Ecology is a scientific discipline but it is also a political philosophy. They are quite different.

The movement spread rapidly across America and to Europe, Japan, Australasia, and eventually throughout the world as a textbook example of the idea whose time had arrived. It was somewhat inchoate, not to say incoherent. Environmentalists sought change for apparently scientific reasons, but few of them were familiar with the science they quoted or even with scientific methods and reasoning. Thus speculation and very provisional research findings were cited as facts, and controversial interpretations of established facts were accepted if they supported a particular opinion. There was much alarm, confusion and exaggeration, leading by around 1970 to warnings that the human race was on the point of self-extermination through overcrowding, pollution, or both, or even that all life on Earth was about to be brought to an end.

The environmentalists sought political reform, and they were influential. 'The environment' became a national political issue and the exaggerations of the early environmentalists amounted to cries so loud they could not be ignored. One by one the governments of the world established departments or ministries to deal with environmental matters.

Then the environment became an international issue, for it was shown that many environmental problems have an international dimension. The seas belong to no nation, and polluted air and continental rivers do not respect frontiers. In 1972 the United Nations convened a Conference on the Human Environment in Stockholm, and several other such conferences were held in succeeding years, each dealing with an issue related to environmental concerns. The Stockholm conference led to the establishment of a new UN agency, the United Nations Environment Programme (UNEP), which is based in Nairobi.

Until Stockholm most environmental issues were identified, publicized and pursued by voluntary groups supported by scientists, but there was little hard research. The scientists involved gave their time, but in most cases they could not commit the resources of the institutions that employed them. After Stockholm the professionals were in command. From the early 1970s environmental problems were referred for investigation to government and university laboratories.

The debate changed, and with it the role of the amateur. It was no longer enough to be vociferous, although that might win popular support. Serious demands for reform must now be supported by hard fact and rigorous argument. Rigorous argument costs nothing, but facts can be difficult and expensive to obtain. They are made difficult by the commercial secrecy surrounding industrial activities, and expensive by the cost of the smallest research programme.

Amateurs have survived, of course, and so have environmentalist campaigns. Since around 1975 the environmental movement has grown increasingly political and political parties or less formal groupings have contested elections. They have had little success in the United States or Britain, mainly because in these countries the electoral system is unable to recognize minority votes, but in West Germany the 'Greens' became powerful during the early 1980s. They command little popular support, seldom winning more than about ten per cent of votes cast, but this is enough to admit them to parliaments, up to and including the federal Bundestag and the European Parliament. The German electoral system tends to produce minority governments which need to form alliances to maintain power, the support of the Greens has been sought to this end, and so the Greens now wield influence that is out of all proportion to the votes they win in elections, although by 1985 there were signs that their share of the vote was declining and their influence waning. Meanwhile, however, the Greens can and do make certain that no German government dares to overlook the issues about which they are most concerned and the German concern is communicated to the other governments with which Germany deals.

As a political force the Greens and related parties in other countries seek radical change. It is difficult to place them on the traditional left-right political spectrum. There are those who see industry as inherently wicked and waste no tears over injuries caused to it, or even at the prospect of its demise. They believe an alternative, but vaguely described, economic system might be more sustainable. The more extreme among them are authoritarian and would impose their ideology without permitting it to be challenged or even questioned. Then there are those who favour a general return to what they suppose would be a simpler, non-industrial way of life organized on something akin to a tribal basis in which government would be mainly by public opinion. This sounds like a recipe for right-wing mob rule. It is confusing and it may be that by entering politics so overtly the environmental movement has lost its sense of direction.

Benefits and costs

Those who campaign for reform often seem to be negative in their approach. This is a sad but unavoidable fact of life, and one of which conservationists and environmentalists are painfully aware. By seeking to repair or prevent damage they oppose much more than they propose and are rarely heard except when they are complaining. They do not like this, for as people they are no more discontented or given to nagging than anyone else, but they cannot help it.

It may be as well, therefore, to recount just a few of the many environmental and conservation improvements that have been made in recent years. Some of them are due directly to pressure brought by conservationists or environmentalists, others were not due to direct pressure but were made because earlier campaigns had raised the general level of public and governmental environmental awareness.

Rivers in most industrial countries remain polluted, but many of them are much cleaner than they were and river and canal banks that were once areas of refuse-strewn dereliction are now pleasant places in which to walk, sit, or even to fish. In the last few years, however, the condition of many British rivers has stopped improving and may be deteriorating due to inadequate funding of anti-pollution measures. British environmentalists are also worried that the privatization of the water industry may lead to a further reduction in water quality. Elsewhere, the

Below *One of Europe's most polluted rivers, the 1,300 km long Rhine rises in the Swiss Alps, flows between France and Germany, then through the Ruhr and the Netherlands.* ***Right*** *A pond filled with garbage, at Estartit, Spain.*

problems that still remain, as with the Rhine and Danube for example, are due to the difficulty in achieving international agreement for rivers that flow through several countries. In time we have good reason to hope those rivers, too, will be clean. It is doubtful that the Danube was ever blue, but one day it will at least be wholesome.

Air pollution has been reduced dramatically, by restricting the burning of coal, by requiring factories to take some steps toward reducing their more harmful discharges, and by imposing stricter standards for vehicle exhausts. The air of industrial cities is not yet clean, but it is much cleaner than it was and more improvements are being implemented.

Many toxic or otherwise harmful substances are controlled more strictly than they were. The addition of lead to petrol will end in a few years. Asbestos can be handled only by licensed operators wearing protective clothing and using approved methods, although the final disposal of old asbestos remains a problem. Radioactive wastes were never a real hazard, to humans or non-humans, but discharges of them are more strictly regulated and are being reduced. Use of most of the pesticides that so concerned Rachel Carson has been much reduced, although some of those old products are still manufactured and exported to developing countries. Yet even this trade has attracted the attention of UNEP as well as of environmentalists and steps are being taken to end it. Modern pesticides are much less harmful environmentally, although some of them are very poisonous to humans.

In the industrial nations wildlife is more strictly protected. In Britain it is now an offence to remove the whole of any wild plant or to harm any wild animal except for certain persons under certain clearly defined circumstances. It is now recognized that agriculture poses the principal threat to wildlife and its habitats, and ways are being sought to modify agricultural practices. Indeed, one of the main criticisms of the Wildlife and Countryside Act was that it required farmers to be compensated for not causing environmental damage. This led some farmers to propose 'improvements' they had never considered before and which they knew would be unpopular in order to claim compensation. Protests

Left *London's river was once little better than a sewer. Now discharges into the Thames have been controlled and it is clean enough to support fish, including salmon.*

Despite its distinguished architecture the 473 MW Battersea power station was coal-fired, releasing polluting dust and gases into the air.

over this abuse masked the main implication of the idea of compensation, which is that farmers were to be paid for managing the countryside other than by farming it intensively, for in addition to agreeing not to carry out an 'improvement' they also had to agree to implement a land management programme prepared by conservation advisers.

If these are the benefits, there have also been costs. There are dangers in demanding too much and some environmentalist campaigns may be harmful. Some countries, most notably the United States, have dealt with the 'environmental impact' of almost every activity by means of regulations that have the power of law. This has led to long and costly lawsuits and also to a sharp rise in industrial costs. It has become so expensive to test a new product to the standards demanded that innovation has slowed.

This is harmful, not only to economic development and employment, but also to the environment. It is generally true that new, more advanced industrial processes are more efficient than those they replace, and this efficiency is achieved in part by the more thrifty use of energy and resources. The thriftier a process the less it wastes, and the less it wastes the less pollution it causes. New processes tend to be cleaner than old ones. If we discourage innovation and continue to use old techniques, and machinery that becomes less efficient and dirtier as it grows older, environmental pollution is likely to increase.

Products may also be affected. The pesticides Rachel Carson criticized were 'broad spectrum' insecticides. They were poisonous to a wide range of insect species and they remained so for a long time in order to go on killing pests. That is why they were harmful. Newer products are much more selective and break down in the environment quickly, leaving harmless breakdown products. They could be made much better. It is possible to develop substances that will kill only certain insects – to kill aphids but not ladybirds, perhaps. Such insecticides would have obvious advantages, but it is now so expensive to test any new pesticide that manufacturers are discouraged. It is safer and more profitable to go on making the older ones – even though they are more harmful environmentally.

How to object

Conservation and environmentalism have acquired political power and sophistication. If local pollution worries you, a development displeases you, a beauty spot is threatened, or any other environmental deterioration provokes you to protest, it is no longer enough to parade with a placard or write to the local paper.

You should contact, or better still join, a locally based conservation group specializing in the subject that most interests you. There are groups devoted to nature conservation, the conservation of buildings, the protection and investigation of archaeological sites, and many more. They are not protest groups as such and that is not their value. Among their members you will find individuals with a great deal of technical knowledge and experience and the group as a whole will have excellent contacts with professionals in their field, in universities or other institutions. The groups are reservoirs of information, and you will need information. After some time you, too, will gain knowledge and experience, and your value to the group will become more than that of a mere subscriber.

Make sure that the problem you perceive really is a problem. It could be that the woodland you believe is to be cleared is of no great scientific, amenity or other value and in any case consists of old, sick trees that will soon die, and the clearance will be followed by appropriate replanting. The pollution that bothers you could be the temporary result of an accident. It should not have happened, but repairs have been made and it will not happen again. Misguided complaints waste time and earn conservationists and environmentalists a bad name. How can you know whether or not your objection is misguided? Make enquiries.

When a genuine problem arises you must define it clearly, and limit it. Your case is weakened if you turn it into a kind of umbrella for your own political or other opinions because you widen your target. If you are critical of a local farmer or a local factory there is no need to attack the entire agricultural industry, or capitalism. They are legitimate targets, no doubt, but it is easier to reform a few individuals than to start a revolution, so start with the individuals!

Before protesting, try using the 'official' channels. There are safeguards and you should start with them. In England, for example, if someone obstructs a public right of way the county council has a legal obligation to remove the obstruction. There is no need to protest – just ask. If water is being polluted the water authority is obliged to deal with the pollution and can prosecute the polluter if necessary. Again, you need only ask. All major changes in land use are likely to be subject to planning laws, and the planning authority – the council – will receive and consider your objection. If you do not know who to approach, ask at the council office or at a Citizens' Advice Bureau. If they cannot tell you, ask members of your conservation group.

If all this fails, then you may need to go further and you will need more information. You may need scientific help, in analysing water or air samples, in evaluating a habitat ecologically, in all sorts of ways. Your group will help and, when you have the information you need, your objection will be much more difficult to ignore because it will be precise.

It is at this point that all else having failed you may need to consider a proper campaign. Do not attempt it alone. You must get help from your conservation group or from local members of one of the larger organizations with experience of planning such campaigns. They may not take up your cause, since they usually restrict themselves to issues with national or international implications, but they will give advice. No campaign deserves to succeed unless it makes the greatest possible use of all the resources available to it. This calls for planning. If you doubt this, watch the way major political parties conduct

Parents and children demonstrating during a visit by Mrs Margaret Thatcher in August 1980. Groups like this, formed to fight local issues, are often very effective, but they need organisation, much hard work, and an ability to raise funds.

election campaigns. The principle is the same.

Part of the campaign will be combined with fund-raising. You should start thinking early about how to finance your activities because the larger your bank balance the greater your chance of success. It is an immoral world we live in, but there it is.

Do not tilt at windmills. There are aspects of modern life you may dislike, but no manner of campaigning will make them disappear. You will be wasting your time and effort. You will not be able to close down any enterprise that is new and in an expanding sector of the economy – though you may well be able to reform it in small but important ways. Factories making micro-processors (chips) in 'Silicon Valley' in California have caused serious pollution of local water through the release of some very toxic substances. The factories can be made to stop discharging toxic effluents, can even be compelled to clean up the mess they have made, but there is not the slightest possibility that environmental protests will close down the factories altogether. The industry is too big and much too important. For the same reason you will not be able to close down a nuclear power station.

Beware of 'backyardism'. It is very common and we are all prone to it, but it is unseemly for all that because it is overtly elitist. It occurs whenever a proposal is made to site a factory, power station, or any other large enterprise almost anywhere. Residents object. They are happy enough to use the products of the factory, and they may boast of their all-electric homes, but they do not want the factory or power station in their own neighbourhood – in their own backyard. How do you avoid backyardism? By being very sure that the site proposed is unsuitable for sound conservation, environmental or other reasons, and not merely for reasons of social prestige. Everyone would like to protect or improve their own backyard, but factories and power stations must go somewhere. What makes you so special?

Campaigns take time. Sometimes they take a great deal of time. They demand commitment, hard work, hard thinking, and cash. You should not launch one lightly and you should accept from the start that you may fail. Then why bother? Because if you organize it well and are lucky you might succeed. Many campaigns have.

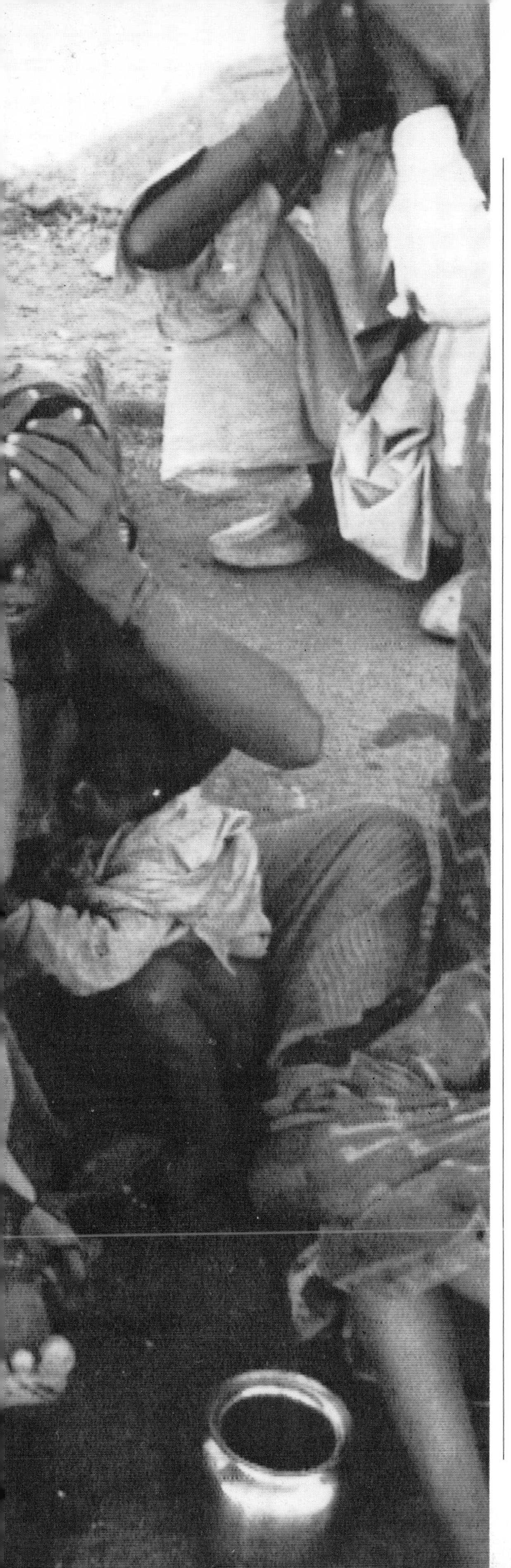

CHAPTER 3

Industry and technology

At about one o'clock on the morning of 3 December 1984 a cloud of gas escaped from a factory about three miles from the centre of Bhopal, in the Indian state of Madhya Pradesh. As people living in the shanty town surrounding the factory slept on, the gas spread, eventually over an area of nearly three square miles. None of the factory workers was affected, but by the time the gas cleared more than 2,500 local residents were dead and more were still dying, and somewhere in the region of 200,000 people had been injured. Some of them may never recover fully. In the following February the *Hindustan Times* reported that nearly 25% of babies born in Bhopal in the two months following the accident died soon after birth, and in the worst affected areas around 30% of babies had a low weight at birth.

It was by far the worst pollution incident in industrial history. The factory was closed at once, but the bitter irony is that it had been engaged in making insecticides that are considered environmentally preferable to the DDT they replace. The Bhopal accident has become the centre of complex legal arguments, but there are environmental lessons to be learnt from the tragedy. The first is that people who decide that certain industrial products are harmful can sometimes form pressure groups which succeed in persuading governments to ban them in favour of alternatives which turn out to be much worse. The second is that safeguards which might work well enough in some places can fail under very different political, economic or social conditions. In America or Europe there would have been no overcrowded shanty town so close to a chemical factory. It would have been easier for the

People at Bhopal, India, on 4 December 1984, one day after the leak of methyl isocyanate that killed more than 2,500 people. Many of the survivors suffered irritation of the eyes and skin, and respiratory difficulty.

Left Victims queueing outside a Bhopal hospital on 6 December are given water to drink as thay await treatment. *Above* Voluntary workers help with eye bathing. Methyl isocyanate can cause ulcers on the eyes, in the worst cases leading to permanent blindness. No more than 30 tonnes of gas escaped from the plant.

emergency services to move in quickly and evacuate the people at risk.

The factory, owned by Union Carbide India, was making sevin and temek, carbamate insecticides that are marketed under the names 'aldicarb' and 'carbaryl'. Officially, sevin is classified as 'hazardous' and temek as 'moderately hazardous' to people using them. They are made, in part, from methyl isocyanate, the gas that escaped. The leak continued for less than an hour and during that time it is estimated that no more than about thirty tonnes of the gas was released.

Methyl isocyanate is heavier than air, so in fairly still air it will sink. It attacks proteins, including haemoglobin and lung tissue, so it can cause respiratory difficulty or failure if inhaled, and it is highly irritating to the skin and to the eyes, where it can cause ulcers to form on the cornea leading to permanent blindness in the worst cases. A concentration of five parts per million will kill half the rats exposed to it (the LD_{50} dose) and in Britain and America occupational safety rules will not permit workers to be exposed to more than 0.02 parts per million, and exposure to the skin is not allowed at all. Because it is highly volatile it is stored as a liquid under pressure and at a temperature of 15°c. If it is exposed to air at a temperature of about 30°c it vaporizes at once. A chemical used as a precursor in the manufacture of methyl isocyanate is best known by its World War I name of phosgene.

In the weeks following the accident several theories were proposed to explain it. Union Carbide, the American parent company, received the reports of the scientists and engineers who worked at the Bhopal plant in the three days following the leak, spent some ten weeks conducting its own scientific inquiry, and conducted more than 500 experiments. In March it announced its conclusions with some confidence. Between about 540 and 640 litres (120–140 gallons) of water entered the storage tank. This started a chemical reaction during which the temperature inside the tank rose to 200°c and the ordinary maximum working pressure increased ninefold. As the temperature passed 38°c a safety valve blew, as it was meant to do in such circumstances to prevent damage to the lining and walls of the tank, although these were corroded. This allowed the methyl isocyanate to enter a second tank containing a vent scrubber. This should have neutralized the methyl isocyanate by allowing it to react with caustic soda, but for some reason the scrubber was overwhelmed.

By the middle of January the factory was closed, the stocks of methyl isocyanate that had not escaped had been neutralized by completing the industrial process and turning them into insecticide, and the people who had fled the city in fear were starting to return.

Seveso and dioxin

Bhopal suffered the worst industrial pollution accident in history, but it was not the only such accident. On 10 July 1976 at the village of Seveso, near Milan in Italy, a factory owned by a subsidiary of the Swiss multinational combine, Hoffmann-La Roche, suffered a similar valve failure followed by the release of a cloud of gas. In this case the results were less serious, but about 500 people showed symptoms of poisoning and 700 people had to be evacuated from the area.

The Seveso factory had been making a herbicide called 2,4,5-trichlorophenoxyacetic acid, better known as 2,4,5-T. As the chemical is constructed from its raw materials it is possible for small amounts of an intermediate product, trichlorophenol, to be converted into tetrachlorodibenzo-*p*-dioxin, which is also known as TCDD or, more popularly, just as dioxin, although strictly speaking the name 'dioxin' refers to a group of compounds of which TCDD is one. The Seveso accident caused the release into the atmosphere of between 2 and 130 kg of TCDD.

TCDD causes severe skin irritation and damages foetuses so that pregnant women who might have been affected were advised to have abortions. Because there might be other effects that have so far not been identified but that might develop years later, an expert on dioxin poisoning advised the Italian authorities that the Seveso villagers should be given regular medical examinations for the rest of their lives. All the domestic animals in the area were slaughtered.

Who are the victims?

These are the most spectacular incidents, but in the world as a whole every year there are dozens of minor accidents in which small amounts of highly poisonous chemicals enter the environment, or are believed to have done so. It is hardly surprising that many ordinary people are alarmed and confused, especially when they learn that the substances involved have impossible or sometimes unpronounceable names like methyl isocyanate or tetrachlorodibenzo-*p*-dioxin, and it is not difficult to fan fears, for these industrial pollutants are highly selective in their choice of victims. They always attack the poorest and most socially and educationally deprived members of the community because it is the poor who live in the shadow of the factory. Some of them work in the factory, but most live there because it is in such industrially degraded surroundings that housing is cheapest.

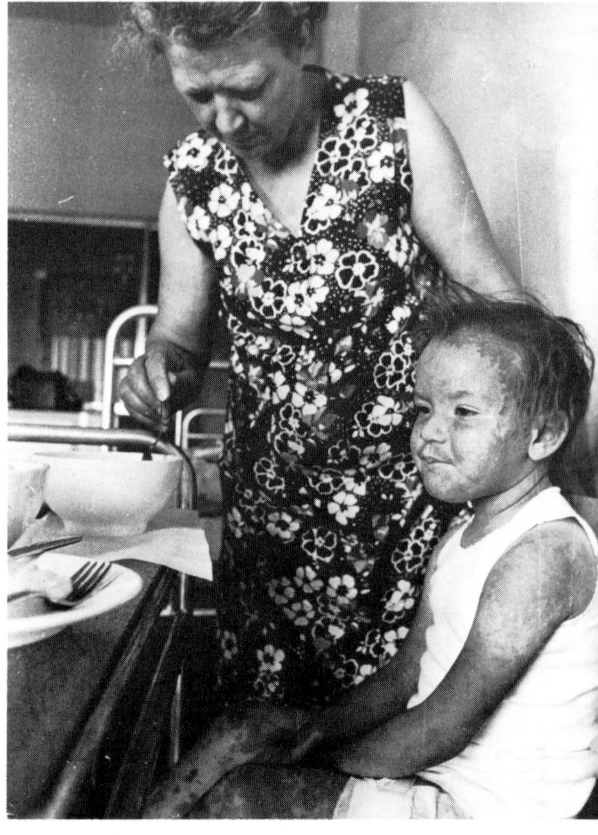

Top *The Icmesa plant at Seveso, Italy.* ***Above*** *No one died from the dioxin leak, but it caused chloracne, a distressing skin condition. Victims included children, like 4-year old Alicia Senno seen here. Dioxin is also suspected of damaging foetuses.* ***Right*** *About 700 people had to be evacuated from the area. A policeman in protective clothing fixes signs prohibiting entry.*

COMUNI MEDA E SEVESO
ZONA INFESTATA
da
sostanze tossiche
DIVIETO
TOCCARE O INGERIRE PRODOTTI ORTOFRUTTICOLI, EVITANDO CONTATTI CON VEGETAZIONE, TERRA E ERBE IN GENERE.
L'UFFICIALE SANITARIO
I SINDACI
DIVIETO DI ACCESSO
ALLE PERSONE NON AUTORIZZATE
I TRASGRESSORI saranno deferiti all'Autorità Giudiziaria ai sensi dell'Art. 650 C.P.

Love Canal

It is not only an active industry that can pose a threat. Poisonous chemicals can appear in the environment years after the process that made them has been abandoned. During the 1940s and 50s an American chemical company, the Hooker Chemical and Plastics Corporation, disposed of its industrial wastes, perfectly legally and properly according to the standards of the time, by sealing them in metal drums and burying them at a place called Love Canal, near Niagara Falls in New York State. By the 1970s some of the drums had corroded so badly that their contents started leaking into nearby water and percolating to the ground surface.

On 7 August 1978, a full generation after the area had ceased to be used for chemical waste disposal, Love Canal was declared a disaster area and 239 families living close to the dump were evacuated from their homes. Later investigations suggested that a further 710 households should be evacuated, but people were reluctant to leave. They were angry and in 1980, when a small sample of them showed chromosome damage, they locked two federal officials in an office for five hours. In the end it turned out that the residents of Love Canal had suffered no serious injury after all. The reports of chromosome damage were not confirmed and there was no evidence of illness that might be linked to such damage. All the same, people had been badly frightened.

Asbestos and lead

In other cases substances that have been in common use for a long time, and whose names are familiar, are suddenly revealed as dangerous. Asbestos, for example, used to be regarded as among the safest of substances. It was so reluctant to react with other substances that it was used to prevent fires spreading, so its name was associated quite directly with safety. It is only in the last few years that we have learned of its dangers when fibres become detached and are inhaled.

Lead was once used widely in paints, and it is only now that paint manufacturers are promising to remove it from their products altogether. It is still used to improve the octane rating of petrol cheaply and it will be some years yet before lead-free petrol is the only petrol you can buy. We have known for many years that lead is poisonous in large doses. It is only recently that we have begun to suspect it may be harmful in very small doses, and that children are particularly at risk because it interferes with the development of the nervous system and brain. Even so, the evidence remains controversial and some scientists are not convinced that very small amounts of lead can cause injury. It is being removed from the environment because governments have been subjected to great popular pressure to remove it and because, despite the scientific uncertainties, they consider it prudent to do so – just in case it really is dangerous.

What makes something poisonous?

Arsenic is a traditional poisoner's weapon, and its name at least is familiar, but today we fear poisoning from elements with more exotic names, like beryllium, cadmium, chromium, cobalt, manganese, and mercury. All of these have industrial uses, and all are, or under certain circumstances can be, dangerous. Chromium was not discovered until 1798 and it has been known to be poisonous since the early 1820s. Most toxicologists would agree that three grammes is usually enough to kill an adult human. Beryllium can cause cancer, for example, and manganese can cause an illness almost identical to Parkinsonism. In the 1950s cadmium caused the condition *itai-itai*, meaning 'ouch-ouch', in the Toyama Prefecture of Japan. Victims, almost all of whom were elderly women who had borne children, suffered deterioration in their bones, causing severe pain in the back and legs. Mercury can cause severe tremor, which used to be known as 'hatter's shakes' because it was common among workers exposed to the metal when it was used in making felt hats. Its victims may also suffer emotional and other psychological changes.

We seem to live in a world fraught with danger. How can we make ourselves more secure? Obviously we should do our best to make sure that industrial accidents are prevented, and that if they do occur the damage they cause is minimized. This is a matter for governments and intergovernmental organizations such as the EEC and the United Nations, which sometimes need prompting from voluntary groups.

Above *Heading for the seaside on a fine summer's day in Britain. As they idle, the cars and motorcycles emit carbon monoxide and carbon dioxide, oxides of nitrogen, the unburnt hydrocarbons that give motor exhausts their characteristic odour, and lead.* ***Left*** *Lead was once used for most plumbing, and asbestos was used widely for packing and insulation and as a construction material. Removing lead is simple and safe, but removing asbestos releases fibres and can expose workers to harmful amounts.*

Japanese victims of mercury poisoning demonstrating outside the Nairobi headquarters of the UN Environment Programme. Between 1953 and 1960, 111 villagers suffered brain damage through eating fish and shellfish contaminated with mercury accumulated from industrial effluent discharged into Minamata Bay.

At best, though, the reforms that result are somewhat piecemeal. They deal with one problem, one pollutant, or one industry at a time. It should be possible to do better, to devise some general principle that might guide us.

We might start where the ecologist starts, with evolutionary theory. Species survive because they are able to exploit, or at least tolerate, the circumstances in which they live. This means that substances which are very common are unlikely to be very poisonous.

Trouble starts when we introduce into the environment substances that do not occur there naturally, or substances that do occur but which we introduce in unusual forms or unusually high concentrations. When our bodies encounter strange substances they may not be able to identify and deal with those among them which are injurious.

Mercury

Mercury, for example, is found everywhere. Tens of thousands of tons of it evaporate each year from the ground surface to be washed down again in rain, so that most rain contains traces of it, but the traces are so small they are close to the very limit of measurement with the most sensitive instruments. Some volcanic eruptions release it. It can occur in its pure form, trickling out of rocks, but most of the mercury used in industry is obtained from an ore called cinnabar, mercuric sulphide (HgS), which can contain as much as 20% mercury, although most cinnabar contains 0.4% mercury or less.

Mercury is common, therefore, but it is spread extremely thinly throughout the environment. High concentrations can occur but they are rare and very local. Humans are not harmed by the amounts to which they are usually exposed. Up to a few hundred parts per billion of mercury is quite harmless. People in Samoa eat tuna fish containing up to 300 parts per billion of mercury, but they are not poisoned by it. It is probable that, like the people living near the shores of the Mediterranean, who also eat fish containing high concentrations of mercury and suffer no ill effects, they are protected because the mercury enters the environment from volcanic eruptions and the volcanoes emit selenium together with the mercury. Selenium is absorbed by the body in the same way as mercury, but is harmless, so that when the two elements occur together they will be absorbed together, and the mercury is diluted.

Alcohol can reduce the harmful effects of inhaling mercury vapour, and the metal itself is not especially poisonous if swallowed because only small amounts of it can pass through the wall of the gut and into the bloodstream. Some compounds of mercury are also difficult to absorb.

Mercury forms the basis of compounds used as fungicides to treat seeds and that are added to paint to prevent the growth of mould on paintwork. Mercuric oxide is used in the small electrical batteries which power torches, transistor radios, small cassette tape recorders and other ordinary, everyday appliances. Mercury is used by dentists in amalgams, in detonators for explosives, and industrially as a cathode in the production of chlorine and caustic soda. A survey once listed more than 3,000 uses for mercury, and every process using it, every product incorporating it, releases a small amount into the environment.

The amount released in this way may amount to some 20,000 tonnes a year over the world as a whole. The amount entering the environment from natural sources is probably more than 30,000 tonnes a year, so the human contribution is considerable. It causes concern because although mercury may enter the environment as the metal itself or as compounds that are fairly harmless in themselves, once there it may take part in reactions, some of which may make it still safer, but some of which may convert it to much more dangerous forms.

What is true for mercury is no less true for the other 'heavy' metals, and for non-metals such as arsenic and asbestos. They are harmless in the forms and concentrations in which they occur naturally, but may become harmful when they are concentrated or changed in form. Cadmium is distributed widely in nature, but rarely in concentrations of more than about 0.2 parts per million. Arsenic occurs almost everywhere. In some places it can be found in high concentrations, of up to several hundred parts per million, but usually the concentration is less than five parts per million. Chromium is an essential nutrient element. We need it. It is only when we take too much that we are poisoned by it.

Man-made poisons

Other substances are man-made and do not occur naturally in the environment, so species have no experience of them and no way of dealing with them if they should prove harmful. They may be harmful if they are similar in most ways to familiar substances, so they can be absorbed. Many modern pesticides fall into this category. They 'trick' their victims into accepting them because they seem familiar, but once absorbed they behave differently from the substances they mimic.

Such poisons sound very alarming, but in fact they are subtle and fairly specific. They represent a considerable advance on the compounds they replace, the 'first generation' of which were much more commonplace poisons, capable of injuring a wide variety of species. Those were replaced by subtler compounds, some of which were intended to be long-lasting, such as the organochlorine group of compounds whose most famous member is dichlorodiphenyltrichloroethane, or DDT. They poisoned insects directly, and once they had been applied they were almost completely insoluble in water and broke down only slowly, so that insects who missed being sprayed might still be poisoned when they returned to the sprayed area, even if they came back a long time later.

These persistent pesticides fell from favour for two reasons. Because they did not disappear quickly from the environment they tended to be concentrated along food chains. The amounts used were insufficient to poison anything larger than the insects at which they were aimed – and humans have eaten large amounts of pure DDT without suffering any ill effect – but an insectivorous bird might consume hundreds of poisoned and probably dying insects. That bird might be caught by a predatory bird, such as a sparrowhawk, and the predator might eat dozens of small birds. The organochlorine insecticides were insoluble in water but they did dissolve in fats, so they accumulated in the bodies of animals which ate them. The insectivorous bird would store in its body the insecticide from all the insects it ate, and the predatory bird would store the insecticide from all the small birds it ate. That is how the insecticide was concentrated and in some cases animals high on food chains, the top carnivores, were harmed. Some predatory birds laid eggs with thin shells which broke, for example, and their populations declined.

At the same time the insects themselves were evolving to resist the poisons. Every time the insecticide was used a few insects would be sprayed but

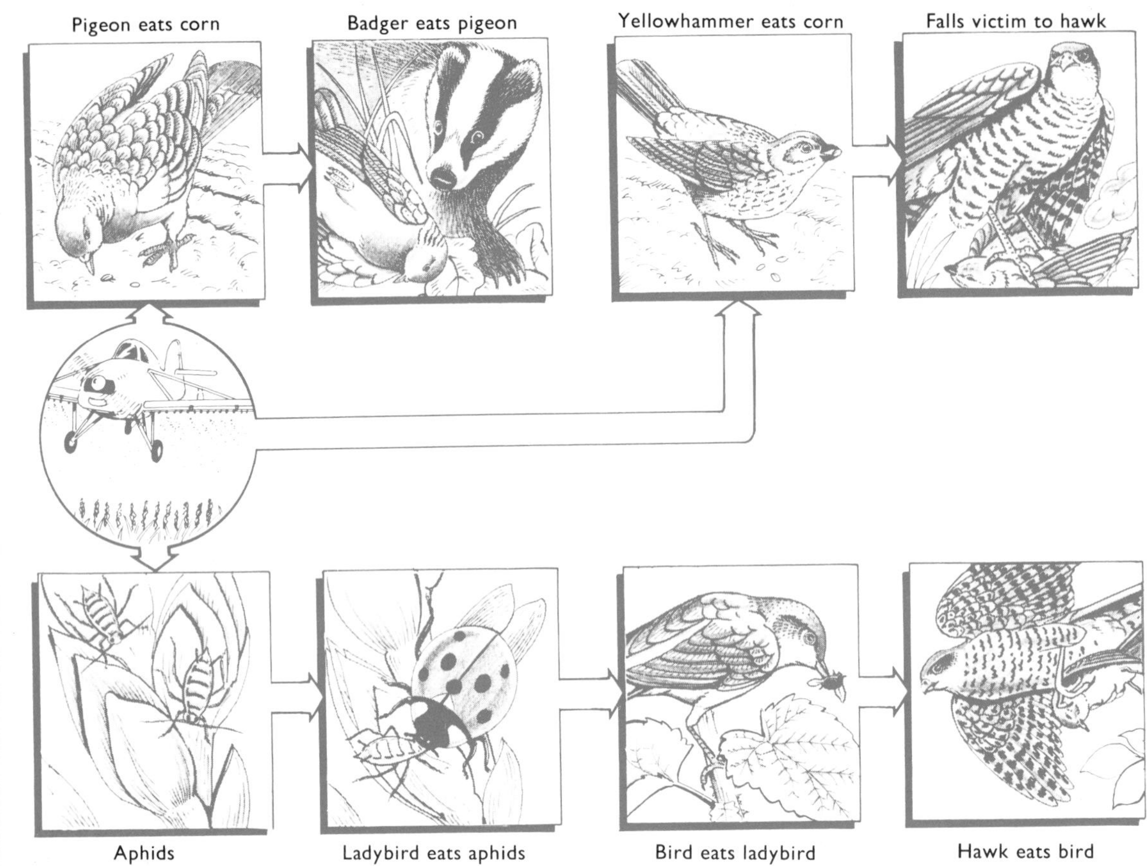

The cumulative effect of persistent pesticides. Some pesticides, especially those of the organochlorine group, are insoluble in water but dissolve readily in fats and oils. When sprayed they will not be washed away by the rain, but an animal eating food contaminated with them will store them in its body fat. If, for example, aphids are sprayed, and each insect stores a quantity '1' of pesticide, the ladybird which eats 10 aphids will receive '10' units of pesticide, and may store all of it. The bird that eats 10 ladybirds will receive a total of '100' pesticide units. In every case of this kind the pesticide dose increases at each stage in the food chain, and top carnivores are likely to be the non-target species most affected.

would survive. Some of them may have had thicker skeletons (an insect has its skeleton on the outside, remember, and the insecticide must pass through it). Others may have had chemical ways to break down the poison into harmless products. These individuals would survive, would be the parents of the following generation and, since their resistance to the insecticide was inherited, they would pass it on to some of their progeny. Little by little the whole population would become resistant.

The new compounds are designed to attack particular target species, hit them hard, and then disappear. They do not concentrate along food chains and they do not remain in the environment to encourage their victims to acquire resistance to them. Some insecticides which chemically are rather similar to nerve gases are extremely poisonous to mammals, including humans, but they are also fairly unstable, and break down rapidly once they have been released, so that in a short time only their harmless products are left.

What can we do to protect ourselves?

It is impossible even to attempt a list of all the industrial products that are, or might be, harmful to humans or to the environment. There are at least 1,000 different pesticide formulations alone. Having decided why it is that certain substances are poisonous, what we can do is divide the pollution problem into two parts: the substances that are released into the environment deliberately, such as pesticides; and those that are meant to be contained but which enter by accident, such as lead. Some substances, such as arsenic and mercury, have to be included under both headings, for both of these are used industrially and in pesticides.

If a substance is meant to be released into the environment and has or is believed to have a useful purpose, then its formulation and use are strictly controlled. There are problems over nomenclature, because many compounds, and pesticides most of all, are manufactured under a bewildering variety of brand names. There are problems, too, arising from the backlog of substances that were in use at the time controls were introduced. This is a diminishing problem, however, mainly because the substances themselves are being replaced by newer ones which have to be registered before they can be marketed. The regulation involves tests to determine their toxicity to organisms they are not meant to harm – including humans, obviously, but not only humans – and tests to trace what happens to them in the kind of environments where they will be used.

In most countries regulations covering the marketing and use of pesticides are compulsory. Until now they have been covered in Britain by a voluntary scheme to which all the manufacturers belonged, but Britain is to introduce legislation to make it compulsory in the near future.

Substances that are placed in the environment deliberately, which are shown to be harmful, and for which suitable alternatives are available, should be removed. This is what usually happens, but it is a great deal more difficult than it may sound. Probably there will be agreement at international level among governments, the oil industry, and the motor industry to abandon the use of lead in petrol, but it will not become fully effective in Europe before the 1990s. Moreover, it will not entirely solve the problem of public exposure to lead because in some areas domestic water supplies are carried through lead pipes and the replacement of that old plumbing is expensive and difficult.

We can agree, perhaps, to stop putting lead into the environment and, if we do get round to replacing

At a filling station in New Jersey, lead-free petrol costs two cents a gallon more than leaded petrol. Lead is unnecessary if petrol is refined to a higher octane rating, or if engines are designed to burn low-octane fuel.

the lead pipes with alkathene ones, scrap lead can be disposed of easily and safely. Asbestos is much more difficult. We can, and over most of the industrialized world do, forbid its use in building, but what can we do about buildings that contain it already? It is dangerous only when its fibres become detached so people can inhale them. Fixed inside roofs or walls, or covered with a generous coat of paint to bind it, it is harmless. It becomes dangerous only when it is removed. The solution, such as it is, is to allow it to be removed or handled only by licensed operators, and a condition for issuing a licence is that the workers be trained properly in the safe handling of asbestos and that they be supplied with protective clothing.

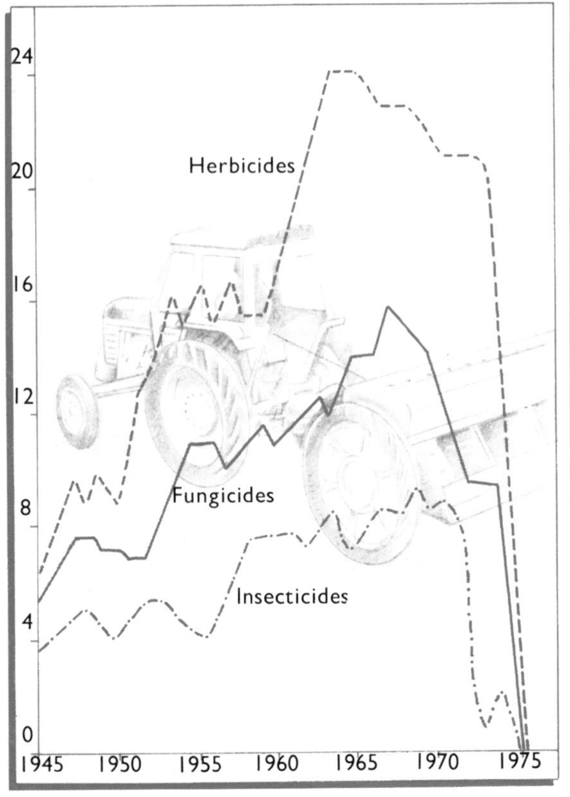

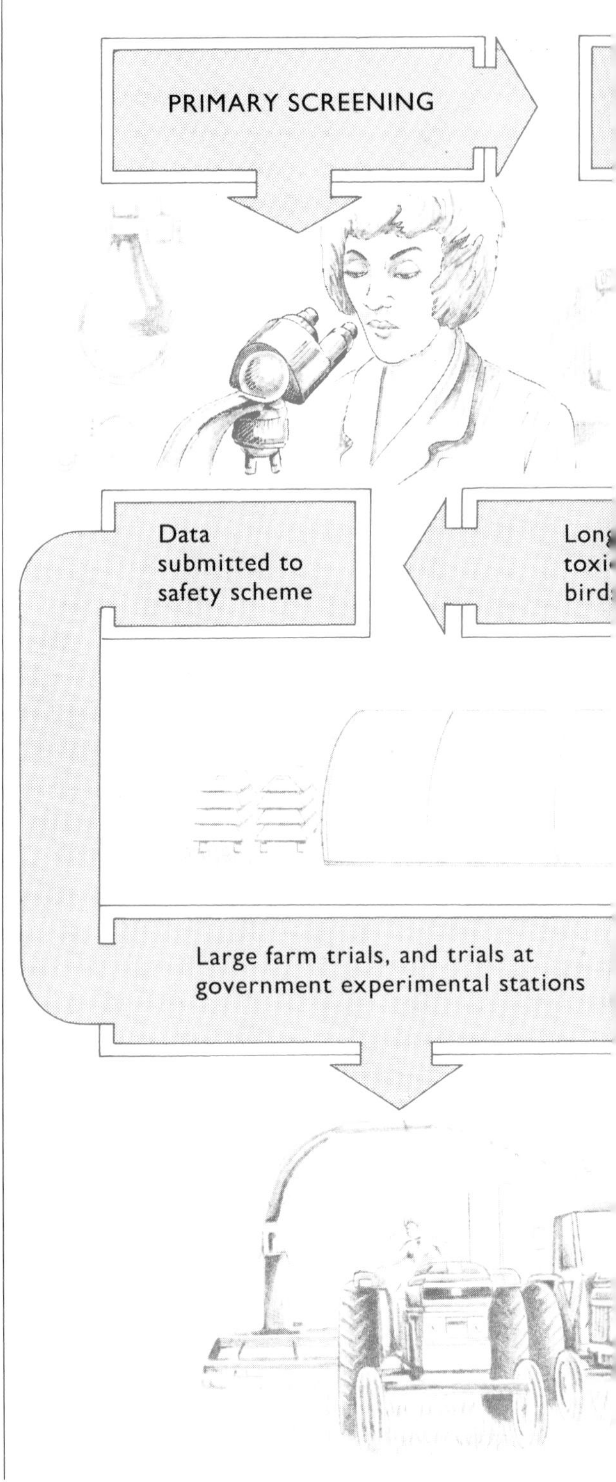

Below *The number (not quantity) of pesticide compounds produced in Britain annually between 1945 and 1975, showing a clear peak in the 1960s followed by a sharp fall as more stringent testing made the introduction of new products more expensive. Today testing for its environmental effects accounts for about half the cost of developing a new pesticide.*
Right *Once a potential new pesticide has been tested for effectiveness it begins a long series of tests to determine its environmental safety.*

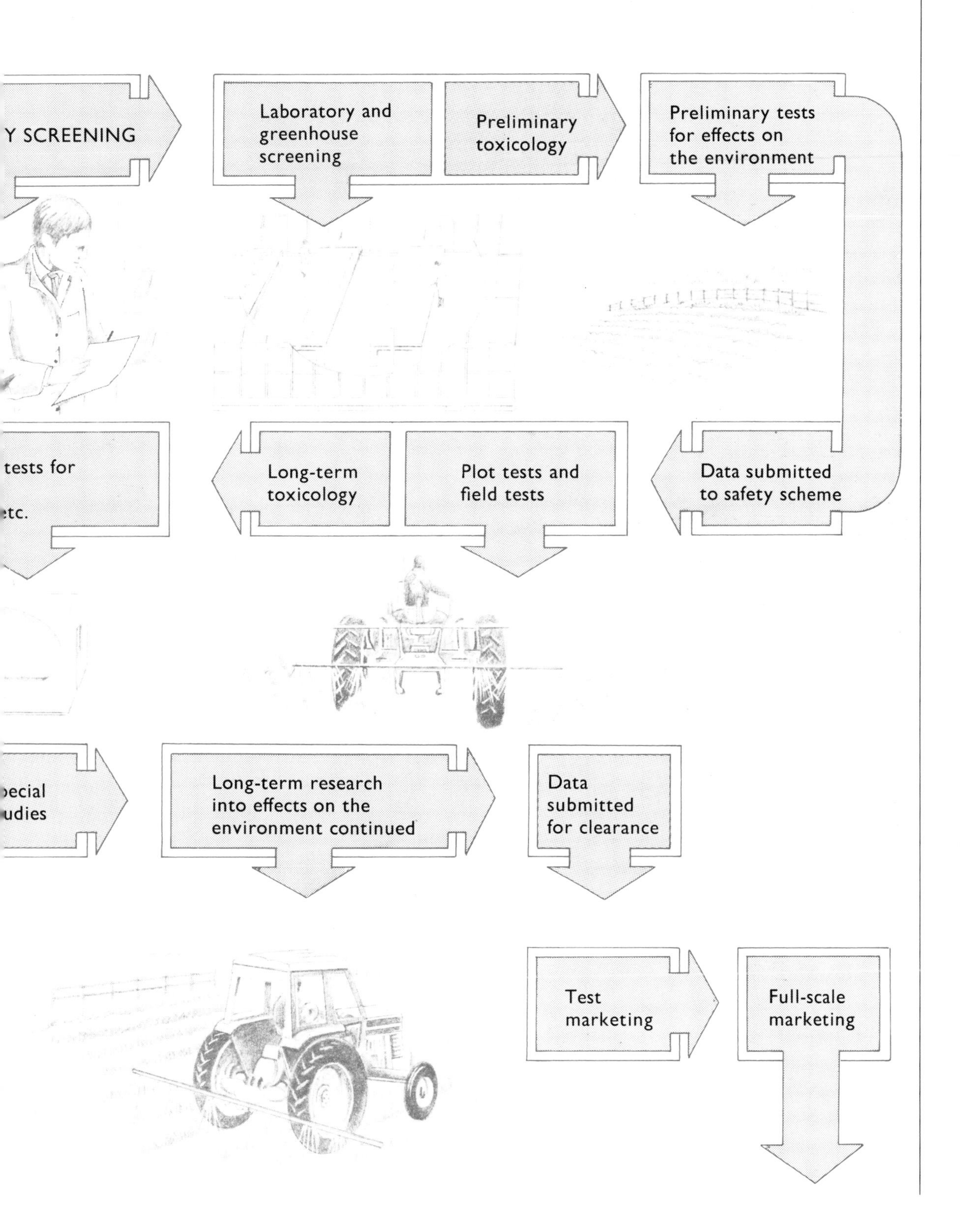
Y SCREENING
Laboratory and greenhouse screening
Preliminary toxicology
Preliminary tests for effects on the environment
tests for
etc.
Long-term toxicology
Plot tests and field tests
Data submitted to safety scheme
ecial
udies
Long-term research into effects on the environment continued
Data submitted for clearance
Test marketing
Full-scale marketing

The use of dangerous chemicals in industry presents quite different problems. Their routine release in effluents can be forbidden, provided the legislation forbidding it can be policed. There is growing acceptance of the principle, adopted officially by the EEC in 1973, that the cost of preventing or remedying environmental pollution should be met by the individual or organization causing it. It has been popularized as the 'polluters pay' principle, although it remains a principle rather than a fact because to some extent care of the environment unavoidably remains a public rather than private responsibility.

There are attempts in Europe, and through the United Nations, to compile a comprehensive register of hazardous industrial substances, the companies making, storing or using them, and to give advance notification of their movement, especially if they are to cross a national frontier. Again, the proliferation of brand names and the tenacity with which commercial secrets are guarded makes the task more difficult than it should be, but progress is being made and one day such a register will exist. It is needed because governments, and their emergency services, often have no idea what substances are being carried along their roads, railways and waterways and should an accident occur, valuable time may be lost while the cause of the accident is diagnosed and appropriate treatment applied. The system of labelling vehicles carrying chemicals is a first step but so far it does not extend over the whole world, and there is a considerable trade in chemicals – and in chemical wastes – between industrialized and non-industrialized countries.

Regulations, limits, assessments

The situation is improving. Attitudes have changed radically over the last twenty years or so. Factories are no longer permitted to pollute the environment as freely as they once were. In most countries the most serious pollutants are identified and rigid limits imposed on the amounts that may be released.

Mining companies must dig for the commodities they extract, but they are no longer permitted simply to abandon sites when they finish working them. The sites must be restored and the plan to restore them is a condition of the permit to open the mine in the first place.

At Pernis, in the Netherlands, the Waterwisser *('Water Wiper') is a prototype vessel designed by Shell to collect floating matter from the surface of water in the petroleum docks. It can be used to remove floating oil from fairly still water anywhere.*

In America no new development may proceed until those proposing it have prepared an 'environmental impact assessment'. This is a careful and detailed study of the likely environmental consequences of the development, together with plans to avoid causing damage or to repair damage that cannot be avoided. Many people urge the introduction of such assessments in Europe.

There are dangers in having too many regulations. The cost of obeying them may inhibit innovation and, apart from the economic effect, this may delay or even prevent the substitution of new products and processes for old, even though the new ones may be less damaging to the environment.

Professionals and amateurs

Protection of the environment has passed into the hands of scientists with the resources to investigate threats thoroughly and governments with the power to act. It is not so simple as it used to be for the amateur to take part in environmental debates. The science is often complicated and open to different interpretations. Yet vigilance is still needed and the voluntary organizations have a useful part to play in drawing attention to possible hazards that may have passed unnoticed and to local abuses of existing regulations.

Left *Brisworthy china clay works, Cornwall. The clay forms deep below the surface and its quality usually improves with depth. Mining creates deep, wide pits, surrounded by flat areas of mica waste and hills of white sand.* ***Below*** *The same works today. The landscape is being reclaimed by pumping a mixture of water and seed, with an organic material to bind it, on to the surface. Trees are planted, and within a few years sheep and other animals graze the pasture.*

CHAPTER 4

The ozone layer

The ozone layer is a region in the Earth's stratosphere where a certain amount of incoming solar radiation is absorbed. Over the last twenty years or so a series of man-made threats to it have been reported and each has been treated with more gravity than the last. Nowadays such threats feature regularly on the agendas of international conferences.

The danger appears to be dramatic. Should the ozone layer be depleted of its ozone the part of the solar spectrum it absorbs, in the ultraviolet wavelengths, would reach the surface. How much would reach the surface depends on the extent of the depletion, but ultraviolet radiation in general and at certain wavelengths in particular is held to be exceedingly harmful. If we lost our ozone protection altogether the ultraviolet radiation streaming to the surface might kill most of the organisms living on dry land, including ourselves. It sounds dramatic, but does such a threat really exist?

What is ultraviolet radiation?

The Sun emits electromagnetic radiation across a wide spectrum. It is most intense at the wavelengths we perceive as visible light, with red light at the long-wave end of the visible spectrum, and violet at the short-wave end. At still longer wavelengths than red light there is infra-red, then heat, then microwaves, TV, and radio waves. At the opposite end beyond visible violet light there is ultraviolet (UV), then X-rays and gamma rays. X- and gamma rays do not penetrate the atmosphere, but UV radiation does.

Altocumulus cloud about 4.5 km above the Sussex Downs. The stratosphere begins at about 10 km, and the ozone layer extends from 10 to about 50 km.

Electromagnetic radiation consists of photons, all of which move at the same speed, although the speed varies according to the medium through which the photons pass. The streams of photons behave in some ways like waves. The difference between one form of radiation and another consists not in their composition or method of propagation, but only in their relative energies, measured either as the distance between one wave peak and the next – the wavelength – or the number of wave peaks that pass a particular point in one second – the frequency. The warm, red glow of the electric fire is not different in kind from sunshine or gamma rays, but only in the amount of energy it has. UV light, with a wavelength between 4 and 400 nanometres (a nanometre is one thousand millionth of a metre), has more energy than visible light.

Light has sufficient energy to penetrate some materials. In the case of visible light we call those materials 'transparent'. UV light can penetrate certain materials that are opaque to visible light, but it penetrates them only weakly. Among the materials it penetrates weakly are the cells of living organisms. If they penetrate only weakly photons are unlikely to emerge again on the other side, so what happens to them? Their energy is transferred to the material they have entered. They are absorbed.

If energy is absorbed we should not be too surprised to discover it produces effects. Some of the effects of UV light are highly advantageous to certain organisms, including humans. In the cells just below the surface of our skin we have substances that react together, using the energy they receive from UV light and produce vitamin D. This is our main source of that vitamin and people who stay too much indoors may suffer from a vitamin D deficiency leading to a softening of the bones. So we need some UV.

Other effects are less beneficial, for within cells exposed to it UV has enough energy to cause breaks in DNA – the compound from which genes are made. Anything that can interfere with DNA may alter the behaviour of cells, killing them or rendering them malignant. It is this that leads many people to regard UV radiation as inherently dangerous. According to this view, the evolution of life on Earth began under water, safely out of reach of UV, because until photosynthesizing bacteria and then plants released free oxygen into the atmosphere, the planet had no ozone layer, and was unprotected. Life was unable to move on to land until the ozone layer had formed.

No one can know whether this description of the early stages in evolution is true, but it leads to certain difficulties. If photosynthesis began deep enough below the surface of water to be beyond the reach of UV, conditions must have been rather dim. Would there have been enough visible light for photosynthesis to occur at all? As soon as about one per cent of the atmosphere consisted of oxygen an ozone layer formed, but how did organisms survive until then? The most likely explanation is that they had ways to protect themselves and could repair any damage they sustained.

It is probable that some organisms existed above the surface of water before an ozone layer formed. There were mats composed of intricate communities of algae, bacteria and cyanobacteria (which used to be known as blue-green algae) whose fossil remains are known as stromatolites. Many scientists believe it was those mats which first released the oxygen that made the formation of an ozone layer possible. Their modern descendants, which seem to have changed little over the thousands of millions of years they have inhabited the planet, shield themselves from UV with a 'skin' of dead cells. Other organisms, living just below the surface of the sea and within reach of UV light, are protected by the presence of nitrogen compounds in the water, which absorb UV.

We protect ourselves, too. It is UV light that makes the skin of pale humans tan in summer. Skin cells are damaged, there may be reddening and swelling, and the skin responds by producing a thicker layer of dead outer cells and by secreting melanin, a dark pigment that absorbs photons without suffering any damage to itself.

If we are exposed to still more UV, and especially UV with a wavelength of about 300 nanometres, it can produce a form of skin cancer. The cancer is not malignant (it does not spread) and can be treated quite easily, but it does occur. However, if the UV is accompanied by visible light, with less energy, cells recover quite well from damage caused by UV. Such cancers do not occur in people whose skins are already dark.

It has been suggested that UV radiation also causes melanoma – a form of skin cancer that is malignant and very dangerous because the incidence of melanomas in humans correlates with the latitude in which the victims live and thus with the intensity of sunlight and therefore exposure to UV. The latitude also determines the temperature, how-

Earth, as seen from Apollo 11, looking at the Pacific Ocean, with north America in the upper right and Australia in the lower left. Analysis of the atmosphere would suggest strongly the presence of life.

ever, and maybe melanomas are more likely to occur in hot climates. Melanomas also occur on the soles of the feet and it is difficult to see how any kind of sunlight could cause them there. The fact is, we do not really know what can cause melanoma and the case against UV is weak.

All the same, our fears of UV continue, just as our addiction to sunbathing does. At one time the UV part of the solar spectrum was divided by wavelengths into three bands, called A, B and C. So convinced are we that the B part of this spectrum is dangerous that it has been renamed. It is now called D, for 'Damaging'. It is the part of the spectrum which causes non-malignant cancers in people with pale skins. Apart from that, no one has ever proved that it actually causes any damage, but there it is.

As a matter of fact we do have some idea of what happens when we are exposed without warning to more UV – nothing. In the hot August of 1972 there was a violent solar storm. It sent out a shower of protons, one effect of which was to strip away about 16 per cent of the ozone layer. It should have increased the amount of UV reaching the surface to well beyond the maximum limit tolerable to living organisms, but apart from the scientists monitoring it no one seemed to notice. In 1895 the volcano Krakatoa erupted, and one effect of that eruption was to destroy almost one-third of the ozone layer. Everything on Earth should have perished, but it survived.

If the intensity of UV-B radiation were to increase it is possible that some plants might be affected. It is not true, as some people have suggested, that all plants would be killed. Indeed, it cannot be true, because in the tropics the intensity of sunlight is such that the surface is exposed to about the same amount of UV as would reach the ground in temperate latitudes if the ozone layer in those latitudes were stripped away completely, yet this region supports the richest, most exuberant life to be found anywhere on the planet.

How does the ozone help?

From about 10 to about 50 km (6–30 miles) above our heads, a very tiny proportion of the oxygen in the air, just a few parts per million of the air as a whole, is different from other oxygen. Atoms of oxygen (chemical symbol O) usually link themselves together in pairs to form molecules (O_2). In this region of the stratosphere, however, a few of them join together in threes, to form O_3, and O_3 is the gas we know as ozone. Sparse though its distribution is, this band in the stratosphere contains about 95 per cent of all the world's atmospheric ozone. It is known as 'the ozone layer'.

Ozone is not stable because a grouping of three is not the preferred arrangement for oxygen atoms. It makes ozone very reactive as it tries to form stable compounds with other substances.

The energy needed to break the bonds holding oxygen atoms together is supplied by UV radiation at the short-wave, or high-energy, end of its spectrum. The atoms re-form, a few of them as ozone, and the ozone also breaks down, either because its molecules receive enough UV energy to break their bonds, or because they react with other substances. More than half the ozone breaks down because of reactions involving compounds of oxygen with hydrogen (HO_X), chlorine (ClO_X) or nitrogen (NO_X). These compounds are even more rare than the ozone itself. There is one molecule of one of them to several thousand molecules of ozone.

So the air contains atomic oxygen (O), ordinary molecular oxygen (O_2), and ozone (O_3), and each time a molecule of oxygen breaks down some UV radiation is absorbed. The ozone layer acts like a

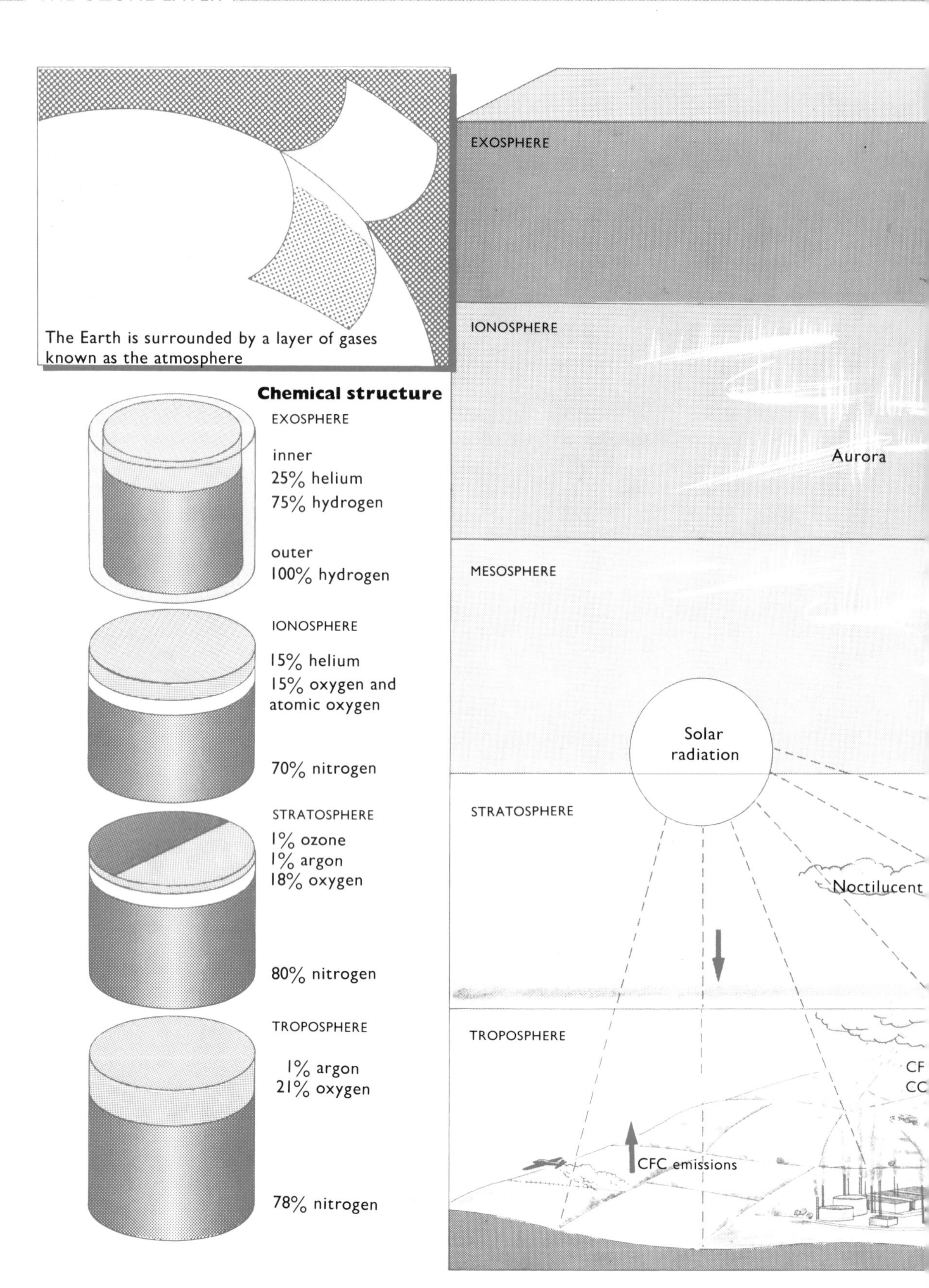
The Earth is surrounded by a layer of gases known as the atmosphere
Chemical structure
EXOSPHERE
inner
25% helium
75% hydrogen
outer
100% hydrogen
IONOSPHERE
15% helium
15% oxygen and atomic oxygen
70% nitrogen
STRATOSPHERE
1% ozone
1% argon
18% oxygen
80% nitrogen
TROPOSPHERE
1% argon
21% oxygen
78% nitrogen
EXOSPHERE
IONOSPHERE
Aurora
MESOSPHERE
Solar radiation
STRATOSPHERE
Noctilucent
TROPOSPHERE
CF
CC
CFC emissions

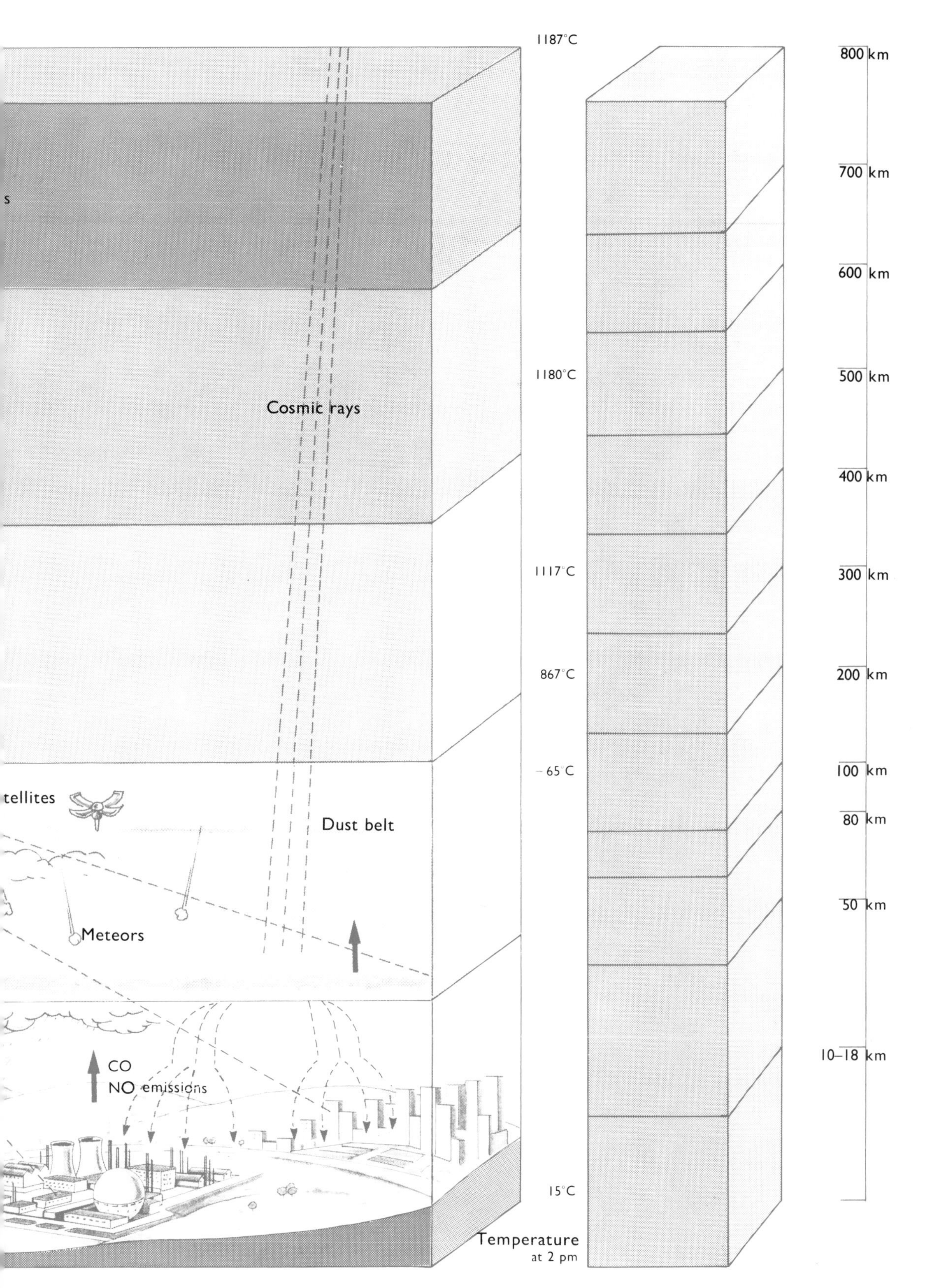
s
tellites
Cosmic rays
Dust belt
Meteors
CO
NO emissions
1187°C
1180°C
1117°C
867°C
−65°C
15°C
Temperature
at 2 pm
800 km
700 km
600 km
500 km
400 km
300 km
200 km
100 km
80 km
50 km
10–18 km

Left *It was suggested that the extensive use of fertilizers containing nitrogen might increase the amount of nitrogen oxides in the lower air, some of which would cross into the stratosphere, where they would take part in chemical reactions that depleted ozone faster than it formed.* ***Above*** *Earlier, it had been feared that supersonic passenger aircraft might release nitrogen oxides directly into the atmospheric ozone layer, so depleting it. Concorde was found to enhance ozone production, however.*

blanket, and the short-wave UV it absorbs is prevented from reaching the surface. The reaction goes on endlessly for as long as the energy is available to power it.

Energy is not available all the time. The energy is sunlight, and at night the formation of ozone ceases, its destruction continues and so the amount of ozone is reduced, only to be restored the following morning. Because conditions in the stratosphere vary from day to day, and the amount of sunlight varies from season to season, the concentration of ozone – the 'thickness' of the 'blanket' – changes constantly. The ozone layer is at its most protective over the poles during their summer, when the hours of sunlight are long, and during the polar winter it almost disappears. Over the equator the layer is more constant, but much thinner than it is at higher latitudes when their hours of daylight are longer. When discussing changes in the ozone layer, therefore, it is advisable to specify how and where those changes are occurring.

What are the threats?

In the late 1960s the Anglo-French Concorde was under development and the United States was also considering building a supersonic airliner. The Soviet Union, too, was believed to be developing a similar aircraft. Supersonic aircraft fly at high altitudes, where the speed of sound is lower and where they can achieve their greatest fuel economy. Until then only military aircraft had flown regularly in the stratosphere.

Like all engines which burn a fossil fuel at a high temperature, aircraft engines produce nitrogen oxides because inside the burners there is sufficient energy to oxidize atmospheric nitrogen. In the stratosphere, nitrogen oxides might react with oxygen and ozone to form stable compounds that would not break down in UV radiation, and this would reduce the amount of ozone. The military aircraft had had little effect because there were few of them, but if three supersonic airliners were to enter service and compete with one another for passengers it seemed possible that before long there would be hundreds of them flying regularly in the ozone layer.

The large fleets of supersonic transports never appeared and so the threat from them receded, but not before Concorde and the Soviet Tu-144 had flown and further calculations had been made. In small amounts, nitrogen oxides enhance the ozone layer rather than depleting it.

The threat receded but the fears did not. New threats were discovered, still from nitrogen oxides. As farmers all over the world spread increasing amounts of nitrogen-based fertilizers on their land, would this not lead to the release of nitrogen oxides from the soil? The oxides might then migrate upward through the lower atmosphere and, although ordinary atmospheric chemical reactions would alter them, some might survive to cross into the stratosphere. Would fertilizer use damage the ozone layer? The theory was examined carefully, and dismissed when it was found that ozone would be depleted much more slowly than it formed.

The dread of ozone depletion caused by nitrogen oxides remained, even then. Every so often, when the risks of thermo-nuclear war are debated, the ozone layer is mentioned. It was mentioned most recently in the context of the so-called 'nuclear winter', the climatic changes some scientists imagine might be caused by such a war. Nuclear explosions release large amounts of energy very locally, and so the fireballs they produce will contain substantial amounts of oxidized nitrogen. Being very hot, the fireballs rise, and so the nitrogen oxides will be deposited in the stratosphere. There is something almost ludicrous in proposing damage to the ozone layer as a threat in any way comparable to the other much more certain consequences of full-scale modern war. Anyway, the warnings might be unnecessary. When nuclear weapons were tested in the atmosphere, and nitrogen oxides produced by the test explosions entered the stratosphere, the result was a thickening of the ozone layer rather than its depletion.

One by one threats have vanished, but then, in 1974, a new one emerged, and it is with us yet: the freons.

The 'Death Sprays'

'Freon' is a trade name, coined by the Du Pont de Nemours company which invented them, for a range of chemical compounds known technically as chlorofluorocarbons, or CFCs. As their name suggests they are made from chlorine (chemical symbol Cl), fluorine (F) and carbon (C) in varying proportions, with or without the addition of hydrogen (H), and they have certain remarkable properties.

They are extremely stable chemically. This means they will not burn. Indeed, they can be used to extinguish fires. They will not decompose or react with any other substance inside the body of an animal. This means they are completely non-toxic.

They are useful because they change from the liquid to gaseous phase at around room temperature, and as they change phase so they absorb or emit heat – the latent heat of evaporation or condensation. Compress the gas a little and it condenses into a liquid, releasing heat as it does so. Release the pressure and it evaporates, absorbing heat. Fill a closed system of pipes with one of the CFCs, pump it around compressing it in one place and allowing it to expand in another and it will cool one area and warm another. This is how it is used in refrigerators, freezers and air conditioners.

This property can be exploited in another way. Seal a CFC compound under pressure in a container that has a one-way valve through which the pressurized contents can be allowed to escape. Under pressure the CFC is a liquid, and occupies relatively little space. Release the pressure and it evaporates at once. This causes a cooling, but it also causes a large expansion – the gas occupies much more space than the liquid. If another substance is mixed with the CFC inside the container, releasing the valve will cause the escape of expanding CFC gas together with the substance mixed with it. This is the principle on which an aerosol can works. The aerosol is the minute particle ejected from the can. CFCs came to be used widely as propellants in aerosol cans.

They are also used to make plastic foams. The foams used as insulation and to make the spongy cushions that pad most modern furniture consist of flexible plastics through which a gas has been bubbled while they were setting. The gas will have been a CFC.

For convenience the freons are given numbers and different freons are used for different purposes. Freon-11 and freon-12 are the ones most commonly used as propellants. Chemically, they are $CFCl_3$ and CF_2Cl_2 respectively. Freon-21 ($CFHCl_2$) and freon-22 (CF_2HCl) are used in refrigeration plants.

CFCs, then, have many uses. The doubts about them arose because of one of their most important virtues – their chemical stability. Once released into the air they tend to remain there. Freon-11 may remain in the air unchanged for as much as 50 years, and freon-12 for up to a century, although there is much uncertainty – for freon-11, for example, the lifetime has been variously estimated as anything from 15 years to infinity.

If they enter air that is warmer than the air surrounding it they will be borne upward. In the lower part of the atmosphere, the 'troposphere', air temperature decreases with increasing altitude. Warm air is less dense than cooler air, so it rises, cooling as it does so. When it is at the same temperature and density as the surrounding air it rises no further. In the stratosphere, which is the atmospheric layer lying immediately above the troposphere, air temperature does not decrease with height, but remains more or less constant. Thus there is a boundary, the 'tropopause' beyond which rising air will cease to rise because it is now at the same temperature and density as the air above it. This makes it difficult for air, and anything carried in air, to move from the troposphere into the stratosphere.

There is not much movement of air across the tropopause, but there is some, and once air – and anything mixed with the air – has crossed the boundary it is even more difficult for it to return. Substances tend to remain in the stratosphere for a long time. If CFCs are released into the lower atmosphere, sooner or later they will start to enter the stratosphere. They are released into the air deliberately, of course, when they are pumped into plastics to make foam, and when they are used as propellants in aerosol cans.

In the stratosphere the CFCs will be exposed to more UV radiation than they encountered at lower altitudes, and UV can decompose them. One of the products of their decomposition is free chlorine, and chlorine atoms will react with ozone. Increase the usually very sparse concentration of chlorine in the stratosphere, therefore, and the rate of ozone destruction will increase. If ozone is destroyed faster than it forms, the ozone layer will be depleted.

Will it happen?

This is the risk and as soon as it was recognized mathematical models were constructed to estimate the long-term effect. The estimates have changed over the years. In 1976 it was predicted that within a

Investigating the stratosphere is difficult. This research balloon will be tracked as it rises into the upper atmosphere carrying a cargo of instruments. When it bursts the instruments will fall by parachute. More than 200 stratospheric chemical reactions are known.

century the ozone layer would be depleted by 6 to 7.5 per cent. In 1979 that figure was increased to 16.5 per cent. Then, in 1982, it was lowered to 5 per cent and in 1984 it was revised again, to 2 to 4 per cent. Now it is changing again, to 10 per cent within 50 or 75 years under certain circumstances. The changes were due to increased understanding of the chemistry of the stratosphere and to changing assumptions about the quantities of CFCs being produced and released.

It is not easy to observe what is happening in the stratosphere. Although it lies only a few kilometres from us, it is extremely inaccessible. Balloons can collect air samples, and remote-sensing telescopes on the ground or on satellites can provide useful information, but that is about the limit. Just a few years ago chemists knew of rather more than 50 chemical reactions that occur in the stratosphere. Today they know of more than 200 – and any or all of them may involve substances that can affect ozone.

These reactions may affect the calculations either way. Some of them increase the amount of ozone. There are also substances in the troposphere that affect the stratosphere without entering it. Carbon dioxide leads to a warming of the air by absorbing heat. The absorption of heat prevents that heat from warming air at a higher level, so if the air does not mix – as it does not across the tropopause – warming in the troposphere may be accompanied by cooling

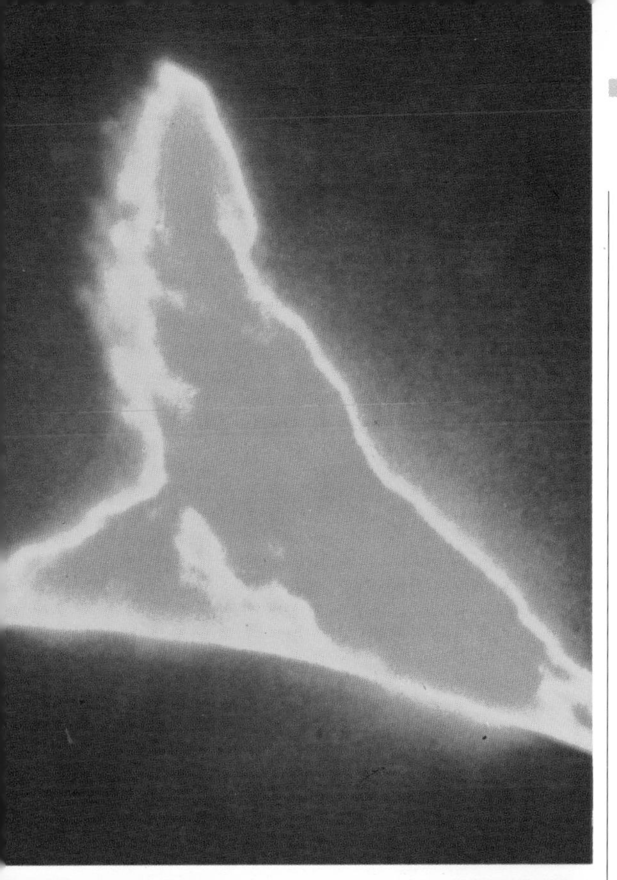

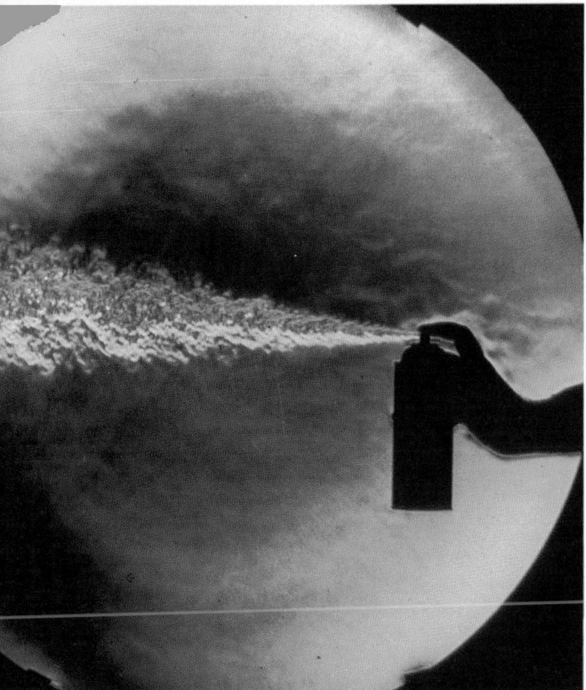

__Left__ A volcanic eruption in Iceland. Each year volcanoes release vast quantities of gases into the air. __Top__ A solar prominence. Storms on the Sun send streams of protons that can disrupt the ozone layer temporarily. __Above__ Aerosols, minute droplets of liquid discharged from a can by the expansion of gas held under pressure. Aerosol cans are almost certainly harmless.

in the stratosphere, and that in turn increases ozone production and decreases ozone destruction.

Nor is it easy to calculate the consequences of the mixing and movement of air in the stratosphere. The largest computers in the world are engaged in work of this kind, for it is the basis of meteorology and climatology, but many of the earlier calculations concerning the ozone layer had to be simplified. They lacked the computing power to do more than regard the ozone layer as constant over the whole planet, for example, so that effects calculated to occur in one place could be assumed to occur everywhere. This was misleading, because the ozone layer is not the same everywhere. It transpired that most of the predicted depletion would occur over the high latitudes during their winter and would have very little effect at lower latitudes.

The rate at which CFCs were being produced was known – at least for the non-socialist countries – and so the rate at which it was being released could be calculated, but it was not known how long the CFCs remained in the air. They do not decompose, but they might be washed out by rain, and they might lodge on solid surfaces with which the air came into contact. If there were 'sinks' for CFCs that removed them from the atmosphere the calculations would have to be revised. Almost certainly there are sinks. It may be that CFC molecules attach themselves to sand grains, so that a proportion of the CFCs released into the air find their way to the world's deserts. Certainly some CFCs must be washed out of the air by rain – but no one knows what proportion. Are CFCs dissolved by the oceans? Such points matter. If a mere two per cent of the CFCs released each year is removed from the air before it reaches the stratosphere the ozone depletion due to freon-11 must be halved and the depletion due to freon-12 reduced to one-third. Monitoring systems are insufficiently sensitive to be able to detect a change of two per cent.

Even then, CFCs account for less than half the chlorine in the stratosphere. Most arrives as methyl chloride, released naturally or as a result of the burning of surface vegetation, or as chloromethane. Chloromethane is produced by many common wood-rotting fungi and chemically it is similar to an industrial CFC. It enters the stratosphere, the bond holding its carbon and chlorine atoms is broken by the energy of UV radiation, free chlorine is released, and free chlorine depletes the ozone layer. The industrial release of CFCs is estimated to be about

A bush fire in Victoria, Australia, emits not only smoke, carbon dioxide, and water vapour, already condensing to form cumulonimbus cloud, but also unburned hydrocarbons and nitrogen oxides. These could react to form ozone in strong sunlight, possibly in the lower stratosphere.

26,000 tonnes a year. The natural release of chloromethane is estimated to be about five million tonnes a year, and it has been going on ever since there have been dead wood and fungi to attack it. If free chlorine depletes the ozone layer, why has the ozone layer survived at all?

What is happening up there?

The concentration of ozone varies from day to day, and sometimes from hour to hour, but by and large the total amount of ozone in the atmosphere has been increasing, mainly in the lower stratosphere. There is about half of one per cent more ozone now than there was fifteen years ago, even allowing for the 'hole'.

The 'hole' in the ozone layer over Antarctica was detected in 1987. Because ozone is formed by the action of solar radiation, but is removed by reacting chemically with other constituents of the air, the ozone layer always thins during the dark winter over both poles. The 'hole' developed between September and November, at the end of the Antarctic winter, and although it filled again quite soon it reappeared in subsequent years, at one point causing a depletion of about 90 per cent in the middle stratosphere and the 'hole' covered much of the continent.

As winter draws to a close, some 20 km above Antarctica air is drawn into a vortex within which temperatures are extremely low and minute ice crystals form in what is usually very dry air. The crystals provide surfaces on which occur the chemical reactions that remove the ozone. Such conditions are unique. A similar but much smaller 'hole' has been reported over the Arctic, with a different chemical mechanism, but this form of depletion is peculiar to polar regions.

What are we doing about it?

Several theories were advanced to account for the unusually large ozone loss but eventually most scientists accepted that chlorine was implicated. It was assumed that CFCs were the source of the chlorine. Press and television reports took up the story. Believing that any depletion in the ozone layer must have dire consequences, suspecting that the polar depletion might spread to cover the entire planet, and never doubting that CFCs were the only possible cause, the reports were needlessly alarmist.

There was a problem, but an indirect one. The solar energy that dissociates oxygen and ozone molecules is absorbed in the process. It is energy that is prevented from reaching the surface and its main effect at the surface is to cause a warming. A serious depletion of the ozone layer might contribute, just a little, to any climatic warming due to the greenhouse effect. What is more, CFCs themselves are very effective greenhouse gases (see Chapter 5).

The popular fear inspired by the press was translated into political pressure. In March 1985, the UN Environment Programme had sponsored the first of many meetings aimed at producing a Convention for the Protection of the Ozone Layer. Several countries, starting with the USA in 1978, had already banned the use of CFCs as propellants in aerosol cans, except for such applications as the spraying of dressings on to wounds and burns, but now other countries were prepared to take action. The Montreal Protocol to the Convention was signed in September 1987. It called for CFC production to be halved, but by the autumn of 1988 many governments were planning the virtual elimination of CFCs.

What are the alternatives?

The production of CFC aerosol propellants, freon-11 and freon-12, reached its peak in 1975 and since then it has been falling. The peak was reached long before any ban was imposed and almost certainly it was due to rising oil prices, which increased the cost of all CFCs and made them rather expensive to use as propellants. They were never used in all aerosol cans, but only in those where the stability of the propellant meant it did not contaminate the substance being sprayed. This is valuable for cosmetics, such as deodorants, hair lacquers and perfumes, but much less important for such products as furniture polish, where it is more advantageous to keep the price low. The recent increases in CFC production have been of refrigerants and gases for making foams. It is these uses that will henceforth be forbidden.

Preferring a total ban to a severe restriction on production, by 1987 leading manufacturers were busy developing new CFCs that will break down fairly quickly in the lower atmosphere. The Montreal Protocol recognised that demand, especially for refrigerants, was likely to increase in Third world countries. This demand will be met by the new products.

Other substances might be used: carbon dioxide would work, so would vinyl chloride, although it is very carcinogenic; and hydrocarbons, such as butane, could be used but are highly inflammable. Ammonia was used as a refrigerant before CFCs were invented.

Costs and benefits

It is true that UV radiation in the B wavelengths is harmful, but its dangers have been much exaggerated. It is also true that the ozone layer absorbs most of the UV-B radiation in the middle and high latitudes. However, if the ozone layer were to disappear completely over the temperate zone the intensity of UV-B would increase only to about the level experienced already in the tropics. There is no reason to suppose the consequences would be catastrophic, although it might be advisable for people to wear hats in summer, and sunbathing sessions would have to be shortened.

It is true that in the stratosphere free chlorine destroys ozone, and that CFCs release chlorine when they are subjected to UV radiation. If we consider it important to prevent the depletion of the ozone layer it may be as well to keep a careful watch on the long-lasting chlorine compounds we release into the air, but we must remember that our industrial releases of chlorine into the stratosphere are minute compared with the chlorine released there naturally – and harmlessly.

The controversy over the fate of the ozone layer has been going on now for many years. It is one of those issues that can be presented in a form sufficiently sensational to attract media attention, leading to pressure on governments for political action, but which is not fully understood scientifically. The action being demanded may be inappropriate or even harmful, as it would be harmful to replace CFCs with substances that really are dangerous, or it may be altogether unnecessary because on closer examination the threat itself evaporates.

CHAPTER 5

The greenhouse effect

For the last four thousand million years our Sun has been growing steadily hotter. There is nothing extraordinary in this. It is typical of stars like the Sun. They burn with increasing intensity until they start to run short of the hydrogen and helium that fuel the thermonuclear reactions by which they are powered. Then they begin the destruction of other elements, such as carbon, which makes them hotter still, until finally they expand rapidly, and then contract and cool. When the Sun expands and becomes a 'red giant', as assuredly it will, Mercury, Venus, Earth and Mars will be consumed as the solar atmosphere reaches out to engulf them. This will not happen for a long time yet. It is another story, a sequel if you like, to the steadily increasing heat to which we and our ancestors have been subjected for so long.

You may be forgiven for having failed to notice the fact that the Sun is growing hotter, because conditions are not growing hotter at the surface of the Earth – at least, not yet. Over the surface of our planet temperatures do not stray much above the freezing temperature of water, or much below it.

The amount of radiation we receive from the Sun is about one-third more than it was four thousand million years ago. If we take our present temperatures and work back, we might conclude that long ago the world was a much colder place than it is today. There must have been much more ice than there is now. However, the geological and fossil records provide no evidence of widespread ice. There have been glaciations, ice ages, but these have always ended and for long periods there has been no permanent ice at all. If, on the other hand, we

Steam rising through the mountain forest from Kamojan geyser, West Java. The vapour condenses to form white steam which reflects sunlight, shielding and so cooling the ground below. The planet regulates its own air temperature.

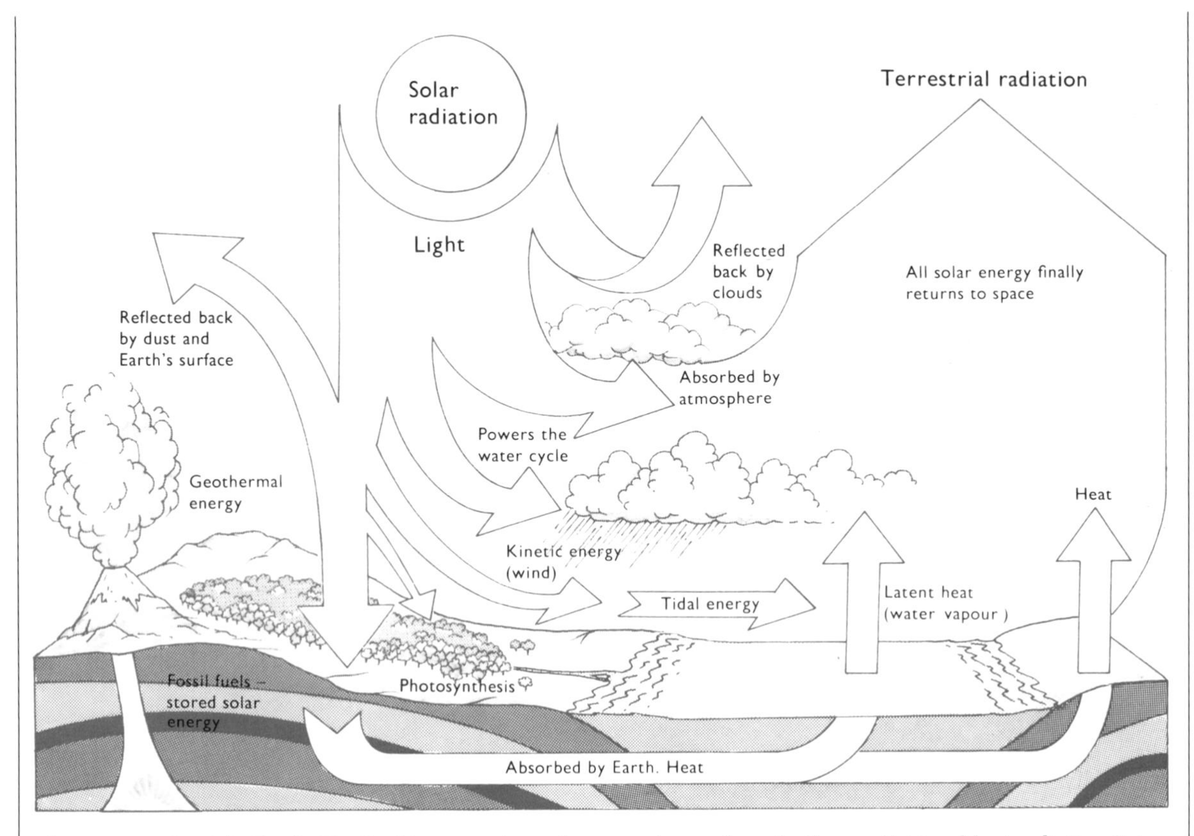

The energy cycle of the Earth. The Earth's core and mantle are hot. Mantle heat warms rocks below the ground surface, and in some places produces subterranean reservoirs of hot water, or hot rocks, which can be used to supply us with geothermal energy. Elsewhere the Earth's own energy may escape spectacularly in volcanic eruptions, or as earthquakes. Gravitational energy drives tidal systems, and the flow of rivers towards the sea. It also affects weather systems and the flow of major ocean currents. Most of the energy used by living organisms reaches us from the Sun, as light and heat radiation. It is solar heat which drives weather systems and causes winds, by warming some areas more strongly than others, and solar heat causes water to evaporate, to form clouds and fall again as precipitation. Sunlight is used by green plants in photosynthesis. Eventually the amount of energy reaching the Earth must equal the amount lost from the surface and atmosphere back into space in order to maintain constant temperatures.

assume that temperatures long ago were mild you can work it out the other way, and conclude that we should all be a good deal warmer than we are now. Then recent glaciations and the present polar icecaps become inexplicable.

Neither calculation makes sense and actually scientists know that temperatures at the surface of the planet have altered very little since life began. A change in the average temperature of just a couple of degrees or so is enough to trigger a glaciation or to end one. This leads us to the big question: how can temperatures remain more or less constant while the amount of heat we receive increases?

The answer, a geologist might reply, is in the rocks. In this case it is rather literally in the rocks, or at any rate in certain kinds of rocks, but to understand what may have happened we must bring the history of our atmosphere into the discussion.

When the planet first formed it had some kind of gaseous envelope surrounding it. As the Sun began to heat, its warmth was enough to sweep away that first atmosphere and for a time Earth had no atmosphere at all. Gradually a new atmosphere was formed from gases released by volcanoes. That second atmosphere contained no oxygen and only a trace of nitrogen, but a very high proportion of carbon dioxide, perhaps amounting to about 98 per cent.

The 'greenhouse' gas

Carbon dioxide is quite colourless. You cannot see it and visible light passes through it as easily as if it were not there. However, all is not as it may seem. When we say a substance is 'transparent' usually we mean that radiation in the visible part of the spectrum passes through it without hindrance – but this is only one part of the spectrum and the substance may be opaque to other wavelengths.

A little more than 40 per cent of the solar radiation we receive has a wavelength between 400 and 700 nanometres (a nanometre, usually abbreviated to nm, is one thousand millionth of a metre) and it is most intense at around 480 nm, the wavelength we see as the colour green. The radiation striking the surface warms it, and the warm surface radiates its heat back into space. The heat radiated from the Earth's surface is radiation, similar to the radiation received from the Sun, but it has a much greater wavelength – from 800 nm up to about 4000 nm.

Although carbon dioxide is transparent to short-wave radiation it absorbs long-wave radiation, essentially any with a wavelength greater than 1000 nm, but it is especially absorptive of radiation with a wavelength of 1200 nm to 1800 nm.

When it absorbs radiation it becomes warmer. This means that energy is transferred from the radiating heat to the molecules of the gas and, having more energy, they move faster. Their faster motion increases the frequency with which molecules collide, and the violence of those collisions. Much of the absorbed energy is spent in warming the surrounding air. However, the warmed gas also radiates some of its heat, and it radiates it in all directions. Some goes up, some down, some sideways, and this re-re-radiated heat warms the ground below and the surrounding air. A large part of it warms other carbon dioxide molecules. So the heat is trapped and goes round and round. The carbon dioxide is like a blanket that holds heat.

As temperatures rise in the lower atmosphere more water evaporates from the oceans, lakes and rivers. Water vapour, too, absorbs radiation, most strongly at wavelengths slightly longer than those absorbed by carbon dioxide. This increases the warming, and it does so fairly rapidly because water vapour absorbs heat even more readily than does carbon dioxide. The blanket becomes much thicker.

This is the 'greenhouse effect'. It earns its name from the rough similarity between the way it works and the way the glass of a greenhouse works. Glass, too, is transparent to short-wave radiation but opaque to long-wave radiation, so the greenhouse traps heat.

If the primitive atmosphere of the Earth was about 98 per cent carbon dioxide it is easy to see how climates might have warmed quite quickly. Had the atmosphere not changed we can also see what would have happened. The weather would have become warmer and warmer. This really did happen on Venus, a planet about the same size as ours but closer to the Sun. Today Venus has an atmosphere which is about 98 per cent carbon dioxide, and the temperature at the surface is about 477°C (900°F). If there is lead on the surface of Venus you should expect to find it molten, flowing in rivers. The atmospheric pressure at the surface of Venus is 90 times that at the surface of Earth. There is virtually no water on Venus. It boiled away into space long ago.

Venus, where the atmosphere is 95% carbon dioxide, 3% nitrogen and 0.5% oxygen. A greenhouse effect has produced a surface temperature of about 477°C, hardly changing between day and night, and a surface pressure about 90 times that on Earth.

The climate of Venus is the product of a greenhouse effect that proceeded to its conclusion and stabilized, so that today the heat received from the Sun is balanced by the heat radiated away from the planet. Earth is further from the Sun than Venus and would not be quite so unpleasant, but had it retained its carbon dioxide atmosphere, the average surface temperature today would be somewhere between 250°C and 350°C (480°F to 660°F) and the surface atmospheric pressure would be about sixty times what it is now.

How do we survive?

The carbon dioxide was pumped into the atmosphere in the first place by volcanoes. To this day volcanic eruptions often release large amounts of it. Today the atmosphere contains only traces of it. Where has it all gone?

Carbon dioxide is weakly soluble in water, so some of it dissolves in the oceans. Quite how much is removed from the air in this way is uncertain, but it could be a large amount. Ocean water is stratified, the warmer water lying on top of colder water, with little mixing of the two. There is some mixing, however, and gradually, over many years, upper water sinks and is replaced by water from below. Surface water saturated with carbon dioxide carries its carbon dioxide below with it and is replaced by unsaturated water. So the absorptive capacity of the oceans could be great.

This was not the main route by which carbon dioxide was removed, however. On land, rocks are heated in summer and chilled in winter, and they crack. Water enters the cracks and freezes, expanding and so cracking them further. Living organisms extract nutrients from them. Little by little the rocks weather away and as they do so the tiny pebbles and fine grains into which they are worn are washed into rivers and carried to the sea. Some of those mineral particles contain silicon, some contain calcium, and they react with dissolved carbon dioxide to form carbonates (CO_3) and bicarbonates (HCO_3). In the seas there are innumerable organisms, most of them so small you need a microscope to see them, which have shells made from calcium carbonate or skeletons formed mainly from silicon compounds.

Most of the time the oceans supply ample amounts of the substances on which these minute organisms feed, but they need raw materials to build their shells and skeletons. They use the carbonates and bicarbonates washed down by the rivers. When they die, the soft parts of their bodies disappear, but their shells and skeletons survive, to be carried by tides and currents as they drift slowly down to the seabed. There they accumulate until, compacted by their own weight, they form new rocks, limestones and chalk. Limestone is defined as a rock containing not less than 50 per cent calcium carbonate ($CaCO_3$), and chalk is almost pure calcium carbonate.

Later, movements of the Earth's crust may thrust above sea level the areas containing those vast layers of rock, bringing them close to the surface on dry land. You can see them today, for example as the famous White Cliffs of Dover, and if you examine them closely, with a magnifying glass, you will see they are made from the shells of tiny marine animals. Limestone and chalk are very common. The world has immense stocks of them. They represent 'buried' carbon, the result of the removal of carbon dioxide from the air.

In other places, at other times, different processes have also removed carbon dioxide. All living organisms are made from chemical compounds based on carbon. The carbon is taken from the air by plants during photosynthesis. Usually, when organisms die their remains decompose, and in the course of their decomposition the carbon they contain is oxidized and returned to the air as carbon dioxide. It simply cycles.

Very occasionally, though, the process of decomposition is halted. It may be, for example, that the bodies of marine organisms fall to the seabed, but before they decompose they are buried beneath other sediments which exclude the organisms that would oxidize their carbon. If other sedimentary rocks then form a solid cap, a cover to seal the half-rotted remains in a kind of natural air-tight vessel, there they will remain. If they are heated they may turn into an evil goo and gases such as methane (CH_4). Methane is what we call natural gas. The goo we call petroleum – a truly classical word made up from the Greek *petra*, meaning rock, and the Latin *oleum*, meaning oil.

A power station near Runcorn in Cheshire, burning coal. When we burn any material containing carbon we release carbon dioxide, and usually some water vapour. Both these are 'greehouse gases' and might contribute to major climatic change.

Or perhaps plants growing in a swamp die, fall, and are buried in thick mud that excludes the air, and thus the oxygen needed for oxidation. Layers of such plant remains, compressed by the weight of sediment above them, heated perhaps, twisted this way and that by movements of the restless Earth beneath them, may turn eventually into coal.

Such processes are very rare. They call for very special conditions. They are also slow. It takes millions upon millions of years to convert dead sea animals into oil, or dead plants into coal. Nevertheless, this is another way in which carbon dioxide, removed from the air, has been prevented from returning to the air.

Taken together, these processes by which carbon dioxide is removed from the atmosphere have proved highly effective. The atmosphere that was once 98 per cent carbon dioxide is now made almost entirely from nitrogen and oxygen. Carbon dioxide accounts for only about 0.03 per cent. Even that small amount is enough to keep the average temperature well above the level it would reach were there no carbon dioxide in the air at all. In that case it would fall to around −25°C (−13°F) and over the whole planet land surfaces would be covered by ice or, at best, tundra.

What happens when you burn fossil fuels

When any compound containing carbon and hydrogen is burned the carbon and hydrogen are oxidized. The hydrogen is turned into water. The carbon is turned into carbon monoxide, which is then oxidized further into carbon dioxide. If the fuel consists of organic material that was buried in the distant past, as part of the carbon-burying process, burning it completes its decomposition and returns to the atmosphere the carbon dioxide it was storing.

In the same way, converting limestone into lime involves heating the calcium carbonate ($CaCO_3$), which drives off carbon dioxide (CO_2) to leave calcium oxide (CaO), which is quicklime, and can be 'slaked' by adding water to it to produce slaked lime ($Ca(OH)_2$). Lime has many uses. One is to absorb the sulphur dioxide produced when fuels containing sulphur are burned.

We have been releasing carbon dioxide into the atmosphere for many centuries, but only recently have we been doing so on a large scale. The first factories of the Industrial Revolution were powered by flowing water, but they soon gave way to factories powered by the burning, first of wood and then of coal, and it was recognized quite soon that the resulting release of carbon dioxide might induce a climatic warming. The eminent British scientist John Tyndall warned of it as early as 1863 in an article called 'On Radiation Through the Earth's Atmosphere'.

Since carbon dioxide traps long-wave radiation it all sounds simple enough. If we release more carbon dioxide, the atmosphere will come to contain more carbon dioxide. If there is more carbon dioxide, temperatures will rise. In the natural environment, however, nothing is simple.

For some years, although scientists agreed that if the concentration of atmospheric carbon dioxide increased temperatures would rise, there was much less certainty about the mathematics of the relationship. No one knew how much carbon dioxide would produce how large a temperature increase. That controversy has now been resolved. The temperature is proportional to the square of the carbon dioxide concentration.

The next puzzle was more difficult to solve. Just because we release carbon dioxide it does not follow necessarily that the gas will accumulate in the air. So will it?

Green plants absorb carbon dioxide. If there is more of it they will grow larger. Horticulturists often exploit this by burning fuel inside their greenhouses in order to increase the concentration of carbon dioxide, and it works. The plants really do grow better. So might an increase in atmospheric carbon dioxide merely lead to more luxuriant plant growth? Indeed it might, but eventually the plants, or the food or other products made from them, will decompose and their carbon will be returned to the air. The organisms engaged in cycling carbon through the environment will have no difficulty adjusting themselves to deal with larger amounts, but those larger amounts will still spend part of their time in the atmosphere.

Perhaps the carbon dioxide will dissolve in the oceans? Perhaps it will – but all of it? No one knows how great the capacity of the oceans may be.

If living organisms have been removing carbon dioxide for so long, and disposing of it safely, can they not step up their operations to cope with the carbon dioxide we release? They are doing so now. Vast amounts of material containing carbon are accumulating on the seabed above the continental shelves and sliding over the edge into the deep oceans. The trouble is that living organisms may be unable to remove carbon dioxide as fast as we are releasing it.

What is happening in the real world?

Although they are based on research, and in some cases on experimentation, the studies of the effects of increasing atmospheric carbon dioxide are largely theoretical. So what is happening in the real world? Is carbon dioxide vanishing into the oceans, into some other 'sink', or is it accumulating?

It is accumulating, but slowly, at about 0.2 per cent a year. The concentration varies from place to place, mainly because the gas is being released principally in the industrial areas of the northern hemisphere and it takes some time for it to diffuse through the atmosphere and be detected at distant sampling stations, such as those in Antarctica. It also varies from season to season, especially at low altitudes. Accurate measurements of carbon dioxide concentration were not kept prior to 1958, but in the late 1960s the 1950 value was calculated at a little more than 300 parts per million by volume and we were then releasing it into the air at around ten thousand million tonnes a year.

For many years all figures earlier than those for 1958 had to be based on estimates, but in 1985 new

The carbon cycle. Carbon dioxide is removed from the air by green plants during photosynthesis, and some carbon dioxide dissolves in water to be used by many invertebrate marine organisms to build their (calcium carbonate) shells. The respiration of living organisms releases some carbon dioxide. As the marine organisms die, their shells may sink to the sea bed, mixed with mud, and eventually be compressed to form sedimentary rocks. When plants and animals die their wastes are decomposed, and the carbon they contain is oxidised and returns to the atmosphere as carbon dioxide. 'Fossil' fuels, comprising organic remains whose decomposition was arrested millions of years ago, release carbon dioxide when they are burned.

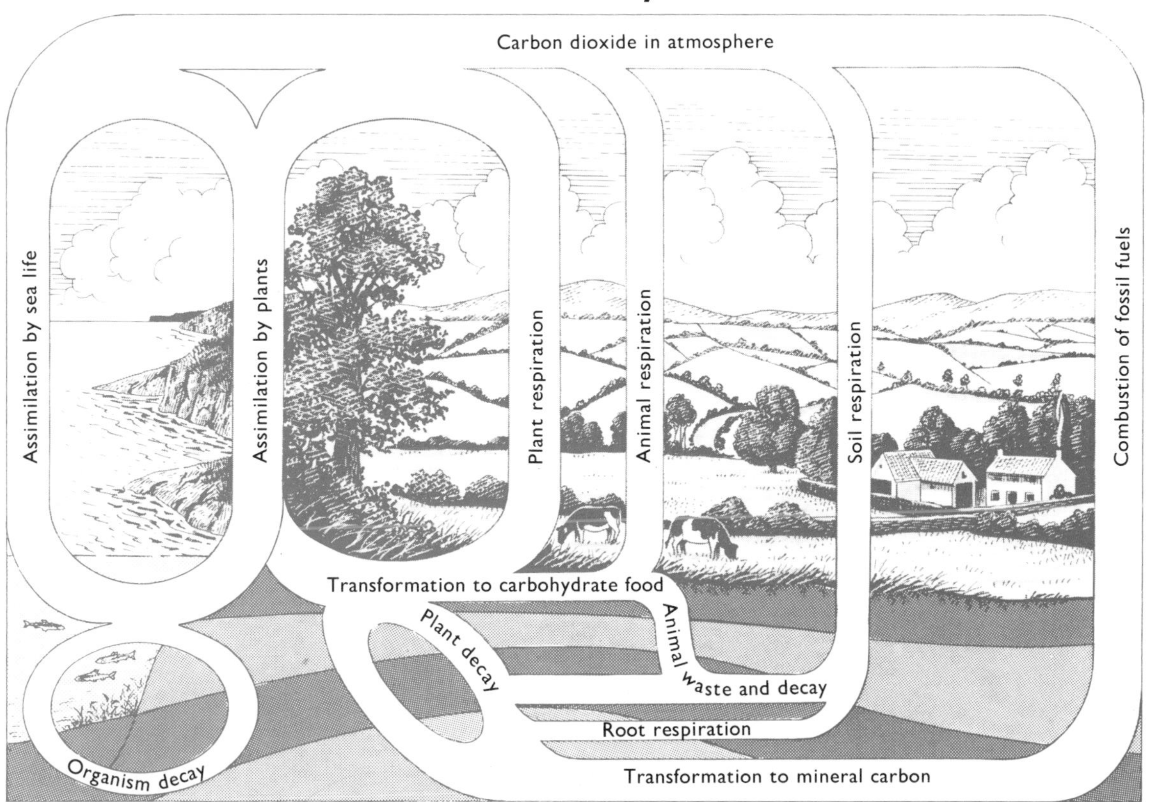

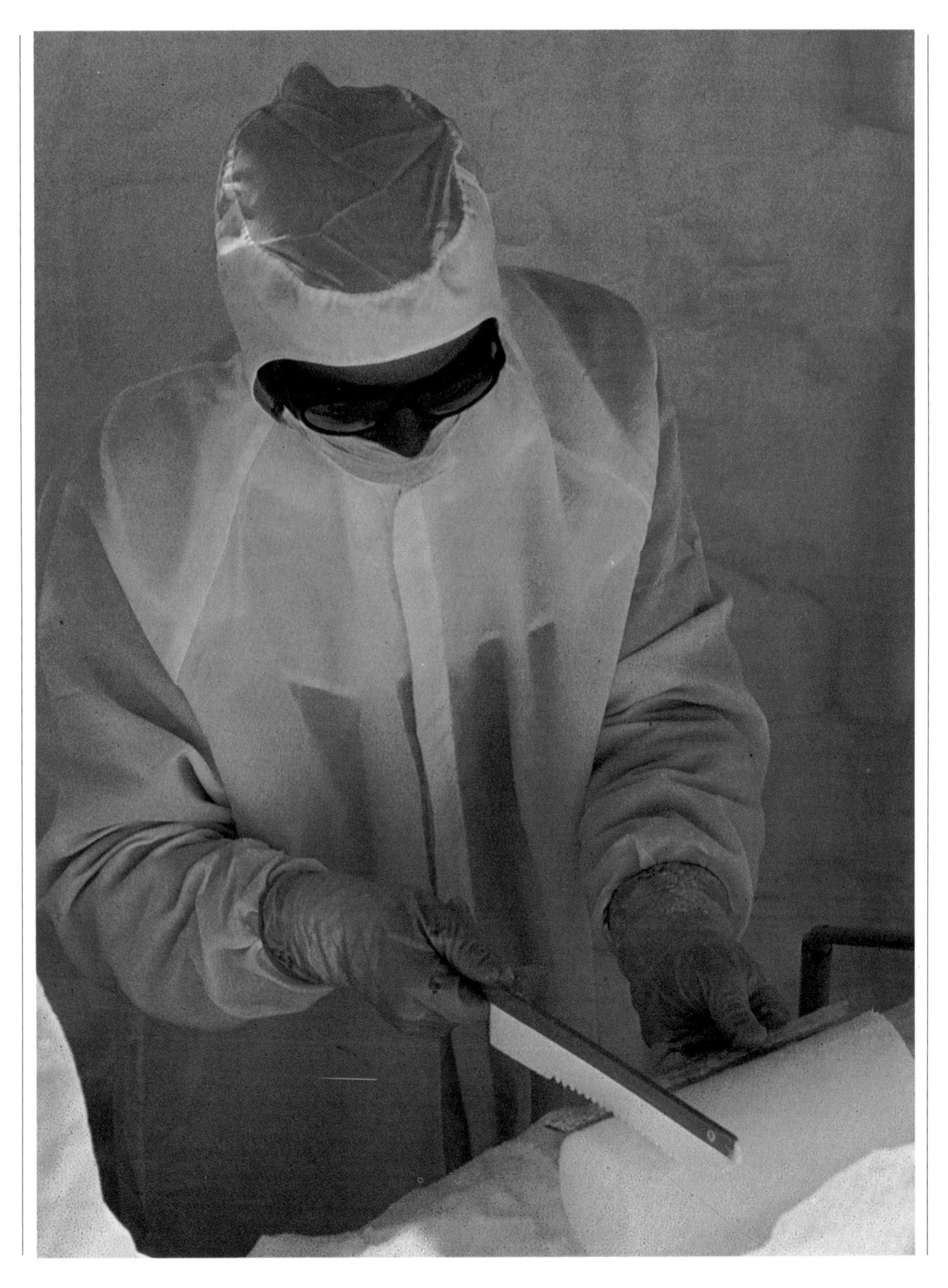

figures were published. These were based on measurement, not calculations. As snow falls on ice sheets it is compacted and hardened by the weight of further falls lying above it, but very small pockets of air remain trapped in the ice. By cutting cores from the ice the trapped air can be reached and analysed. The air has been isolated from the atmosphere since the ice was formed, and because the rate at which ice forms is known it is possible to date the samples. These figures showed that for centuries before the start of the Industrial Revolution the concentration remained fairly constant, but that it has increased recently. In the fifteenth century the air contained about 270 parts per million by volume (ppmv) of carbon dioxide, by around 1750 it contained around 265 ppmv, at about the beginning of the twentieth century it contained about 270 ppmv, and in 1984 it contained 345 parts per million.

There is no longer any doubt at all. The carbon dioxide we are releasing into the atmosphere is accumulating.

Other chemicals, manufactured industrially and released into the atmosphere accidentally or in some cases deliberately, also absorb long-wave radiation. Chlorofluorocarbons, the CFCs that are accused of threatening the ozone layer in the stratosphere, for example, absorb radiation one thousand times more strongly than does carbon dioxide. Their concentration in the atmosphere is very low, but multiply it by one thousand to its 'carbon dioxide equivalent' and CFCs may add significantly to the warming.

Other industrial chemicals also add to it and so do certain other gases including nitrous oxide (N_2O) and methane (CH_4). Nitrous oxide is released by bacterial action from soils, especially in the humid tropics, as a result of fertilizer use, by many factories and especially power stations, and in vehicle exhausts. Methane is released mainly by bacteria in the digestive systems of termites, but also in the guts of farm cattle. Nitrous oxide and methane are particularly significant because they absorb radiation at wavelengths between 700 and 1300 nm. Carbon dioxide and water vapour do not absorb radiation strongly in this waveband and so previously it was thought to be a 'window' through which radiation could escape.

Ice sheets form as snow is packed hard by the weight of new snow above. Small pockets of air trapped in it can be taken and analysed from cores drilled from the ice, and the level from which they came can be dated, so revealing atmospheric composition in the past.

Does it affect the weather?

If the lower atmosphere becomes warmer, more water will evaporate. This will increase the warming, but it will also have another, opposite effect. There will be an increase in the amount of cloud. Cloud reflects back incoming radiation, and casts shade on the surface. This produces a cooling.

The climate changes naturally over the years. At present we live in an interglacial, between two ice ages. In the middle ages, and until the last century, there was a cold period, sometimes called the little ice age. It is possible that the natural trend at present is for the little ice age to return, for the climate to grow cooler. If that is so, the greenhouse effect must overcome it before it can lead to any warming.

There are many other influences on weather during short periods of a year or two. Large volcanic eruptions that eject dust particles into the stratosphere have a cooling effect. The dust reflects back incoming radiation, but the extent to which it will do so is very difficult to predict or even to measure. In recent years Mount St Helens and the Mexican volcano El Chichón have erupted. Mount St Helens produced almost no climatic effect, but El Chichón did, although it was small.

It is very difficult indeed to detect small changes in the average temperature over an entire hemisphere, yet those are the changes that will indicate an overall climatic change.

With all these qualifications to make any statement much less certain than it may sound, climatologists believe the northern hemisphere is growing warmer. So far the change is small, a fraction of a degree in the average temperature, but it has been detected. Is it caused by the greenhouse effect? No one can say, but the extent of the warming is about what calculations of the consequences of the greenhouse effect predict. Will it be sustained, or will the world grow cooler again? No one knows. However, the calculations and measurements that have been made, and the slight warming that has been observed, mean we should take the greenhouse effect seriously. Most climatologists now agree that we should expect major changes in climate during the next fifty to one hundred years.

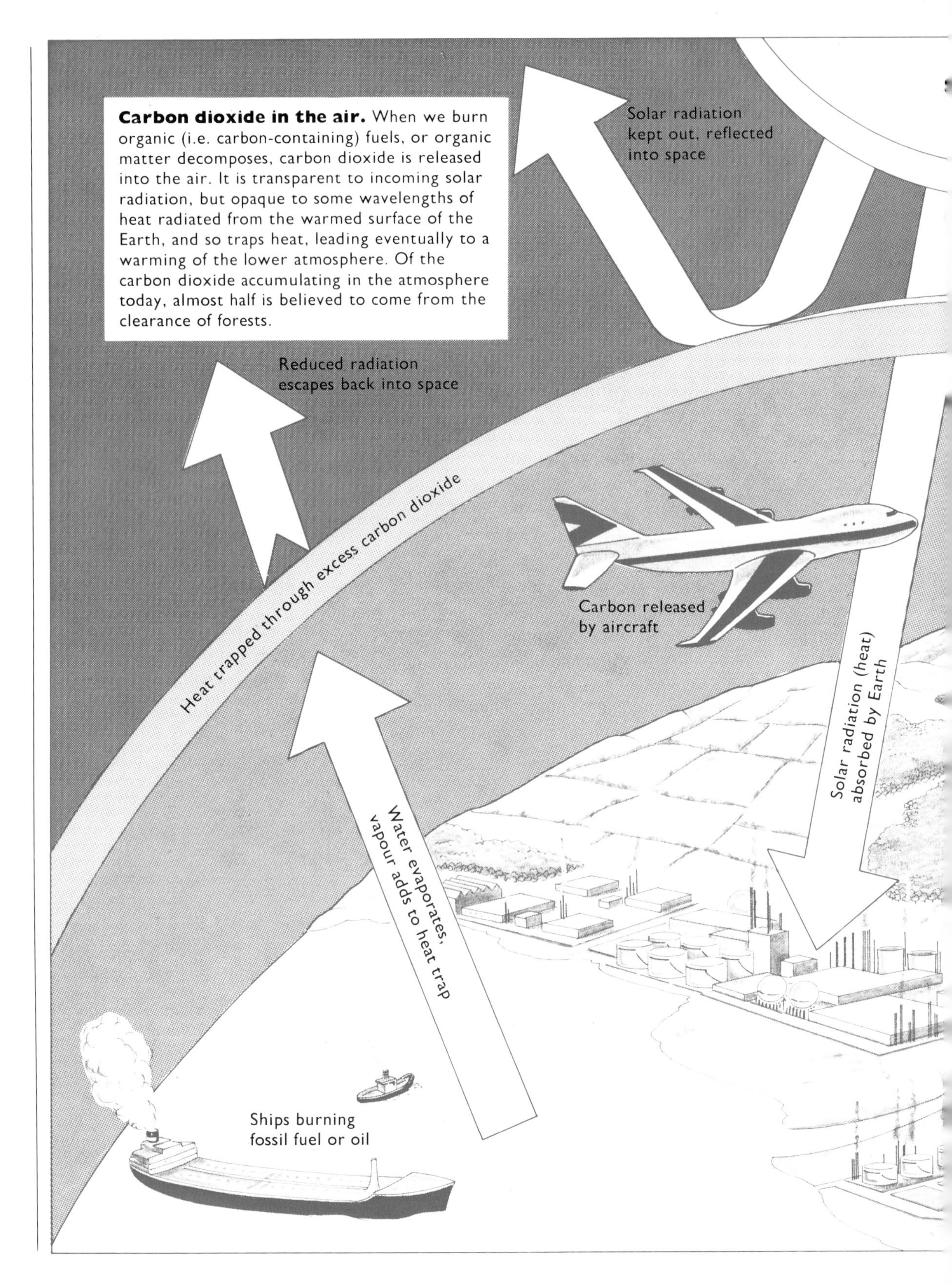
Carbon dioxide in the air. When we burn organic (i.e. carbon-containing) fuels, or organic matter decomposes, carbon dioxide is released into the air. It is transparent to incoming solar radiation, but opaque to some wavelengths of heat radiated from the warmed surface of the Earth, and so traps heat, leading eventually to a warming of the lower atmosphere. Of the carbon dioxide accumulating in the atmosphere today, almost half is believed to come from the clearance of forests.
Solar radiation kept out, reflected into space
Reduced radiation escapes back into space
Heat trapped through excess carbon dioxide
Carbon released by aircraft
Solar radiation (heat) absorbed by Earth
Water evaporates, vapour adds to heat trap
Ships burning fossil fuel or oil

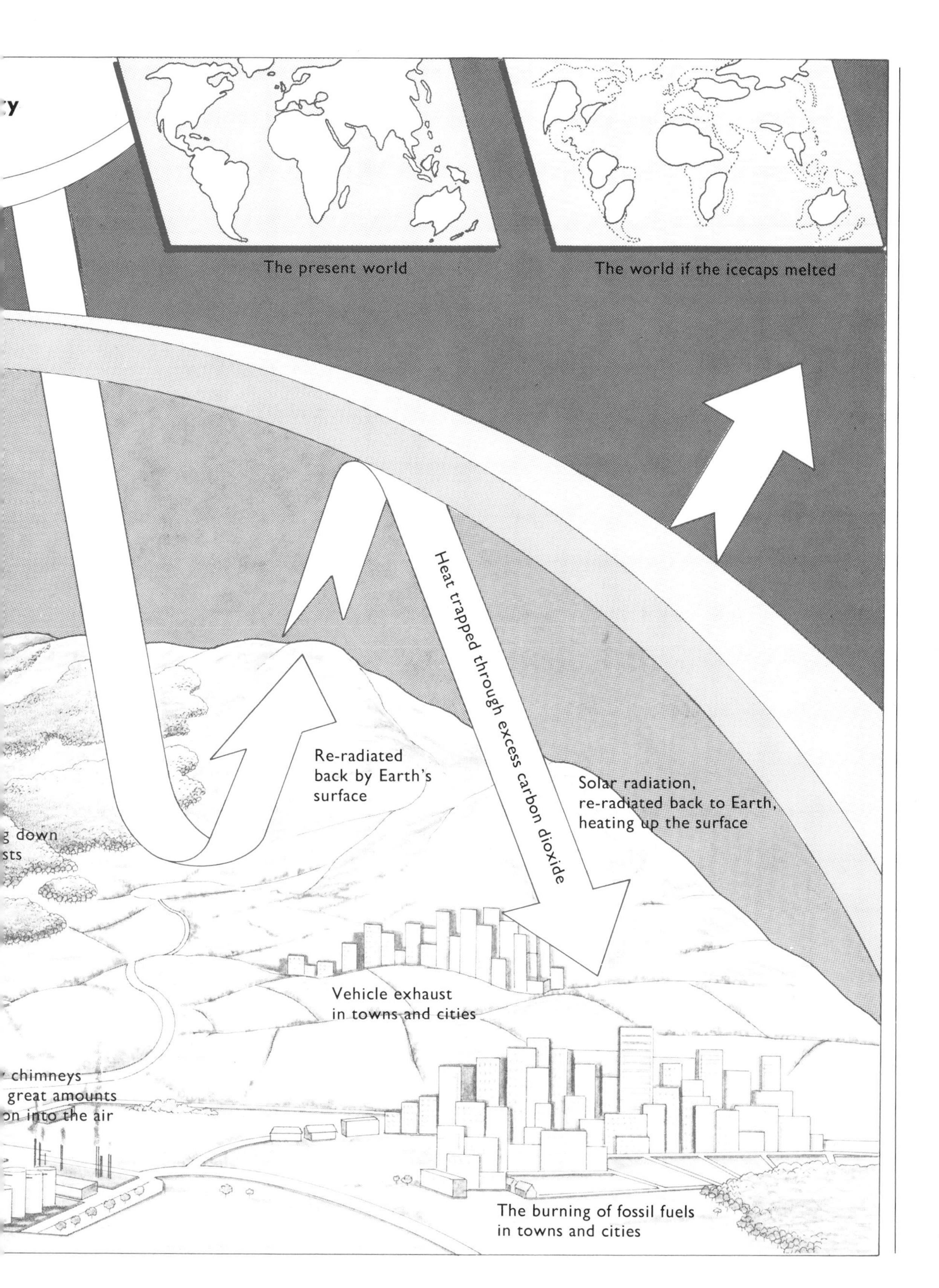
The present world
The world if the icecaps melted
Heat trapped through excess carbon dioxide
Re-radiated
back by Earth's
surface
Solar radiation,
re-radiated back to Earth,
heating up the surface
g down
sts
Vehicle exhaust
in towns and cities
chimneys
great amounts
on into the air
The burning of fossil fuels
in towns and cities

__Left__ In strong sunlight unburned hydrocarbons and other air pollutants react to form photochemical smog, seen here in Montreal. __Top__ The Antarctic ice sheet contains about 90% of the world's ice, some 30 million cubic kilometres. That is sufficient water to raise sea levels significantly were it to melt. __Above__ If the climate warmed, continental interiors might be even drier and these sparse grasses would disappear from the Namibian desert.

The Water Cycle

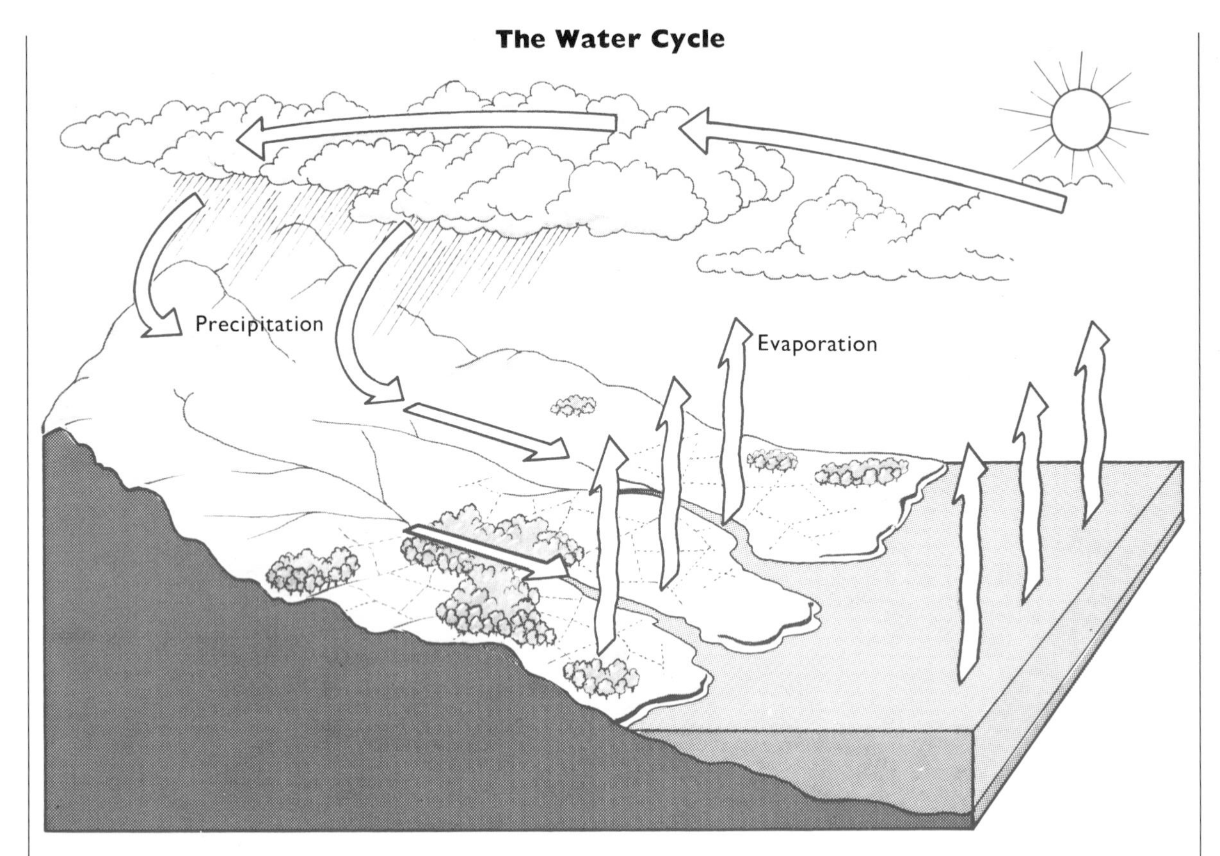

The water cycle of the Earth. Every day about 875.3 cubic kilometres of water evaporate from the oceans, and about 160.5 cu.km. from the land. 100 cu.km. of water vapour is carried from the oceans across the land, 775.3 cu.km. falls as rain over the oceans, 260.5 cm.km. over the land, and 100 cu.km. flows from the land to the oceans through rivers. At the top of the atmosphere a very small amount of water vapour is dissociated into its constituent hydrogen and oxygen and the hydrogen is lost. The loss is made good by 'juvenile' water released from below ground by volcanic eruptions.

Will we all fry?

The effects of a slight warming are likely to be subtle. That is not to say they may not be very dramatic, but only that they will be much more complicated than a simple warming. The greenhouse effect will not necessarily bring us mild winters and splendid summers.

There will be an increase in the evaporation of water, and therefore an increase in rainfall, but between the time it evaporates and the time it falls as rain, water vapour is transported, sometimes for long distances. In the interior of continents it is possible that water will evaporate from rivers and lakes, and that when rain does fall much of it will evaporate quickly from the ground surface. The water vapour may then be carried away, to form cloud and rain over distant mountains, or over the sea. With less water on the ground, lakes may shrink, river flows may decrease, and continental interiors may become more arid. With more water vapour being transported, however, places with oceanic climates – such as Britain – may become wetter while for much of North America the weather could become drier, and both from the same cause.

Changes in rainfall will be associated with changes in temperature because the amount of cloud affects the temperature below the cloud and because water tends to absorb heat slowly and release it again slowly, so it moderates temperature changes. Where rainfall increases the weather may be generally cooler, but in areas that become drier extremes of temperature may be more common. Extremely hot, dry summers and extremely cold winters may occur more often.

The effects may be influenced by the behaviour of plants. They grow more strongly when there is more carbon dioxide, partly by closing their stomata – the 'pores' on their leaves through which water evaporates – and using water more efficiently. This means they remove less water from the ground, which may partly offset the loss of water from increased evaporation. Even so there are many uncertainties. Will plants as a whole grow more abundantly because of increased warmth and carbon dioxide, or less abundantly because of reduced rainfall? Will they grow more abundantly in some places, making wet areas still wetter, but less abundantly in dry places? No one really knows.

A change in our weather could be very important. Our crop plants have been bred carefully over centuries to thrive in the climates in which we grow them. If the main cereal-growing areas, in North America, Europe, the USSR and Australia became much drier, their productivity would fall. In Britain, which also grows cereals, productivity would also fall if conditions became wetter. So major climate change might have very serious consequences for the world's food supply if farming systems were unable to adapt – and at present they are robust and fairly adaptable. Of course, regions that at present have climates unfavourable for agriculture might improve, and this would compensate us, but with unsettling political and economic consequences.

Everything depends on the extent of the change. The US Environmental Protection Agency predicted in 1984 that there might be a rise in the average temperature of 2°C (3.6°F) by the middle of the 21st century and a rise of 5°C (9°F) by 2100. At about the same time the US National Research Council produced its own report, with similar predictions but delayed by about 25 years. A 1°C (1.8°F) warming would have little effect, but a 4°C (7.6°F) rise could be catastrophic, especially in the south-western United States.

Eventually, sufficient warming – and 4°C might be enough – could produce the effect everyone fears. The Antarctic ice sheet could begin to melt. If this happened – to the largest area of snow and ice on the planet – it might turn an area that at present is white into one that is very dark. Because light colours reflect radiation and dark colours absorb it, if this happened it could accelerate, as the dark surface absorbed heat, warmed, melted more snow and ice, exposed more dark areas, which also warmed, until the entire ice sheet disappeared. If the southern ice cap melted, the smaller northern one would also melt. The polar ice caps contain about 98 per cent of all the world's fresh water, and it would enter the oceans. Sea levels would rise by about 50 metres (165 feet), inundating low-lying land areas that include many of the world's major cities.

What can we do?

We can do nothing about releases of carbon dioxide that have taken place in the past, and we still know too little about the greenhouse effect to be confident about predicting its consequences. It may be that the warming trend will be reversed and we shall need to take no action.

If action does prove necessary it will be drastic. It is possible to remove carbon dioxide from the products of combustion before it can be released into the atmosphere, but there is no gain. It would have to be removed by passing the exhaust gases through solutions of lime-water – water containing calcium hydroxide. The carbon dioxide would react with the calcium hydroxide to form calcium carbonate and water – a process that would be a kind of industrial equivalent of the natural process by which so much carbon dioxide has been removed from the air in the past. The gain vanishes, though, when we have to obtain the calcium hydroxide from calcium carbonate (limestone and chalk) by heating it to drive off carbon dioxide. We would simply add carbon dioxide to the air in one place in order to remove a similar amount from the air in another place.

The only practicable way to prevent carbon dioxide entering the air is to burn less carbon-containing fuel. We will need to burn less wood, less oil, less peat, and less coal. We will need to rely much more heavily than we do now – or even than we plan to do in the near future – on nuclear power. Nuclear power generates heat without any production of carbon dioxide. If the greenhouse effect emerges as a real and immediate problem, as it may, nuclear fuels will be the only ones permitted to us, but if you feel you should start a campaign to limit the greenhouse effect this is the line you will have to take. At present there is no such campaign. Perhaps there should be.

CHAPTER 6

The humid tropics

Most ecologists agree that the great forests of the humid tropics are being cleared at an alarming rate, and that this clearance is undesirable. Some scientists would go so far as to say that it is dangerous – for one thing there may be a link between forest clearance in the tropics and the greenhouse effect.

Carbon is the element that links the humid tropics to the greenhouse effect. When plants die and decompose the carbon they contain is oxidized and returns to the air as carbon dioxide, but it does not all return at the same rate. The roots of the plants decompose below ground level and the carbon dioxide produced by their decomposition remains trapped in small pockets of air between soil particles, so that the air held in the soil is very much richer in carbon dioxide than the free air above the ground.

Usually we do not see plant roots, so it is easy to forget about them, and to think of the whole plant as just the part of it we can see. It can come as a surprise, therefore, to learn how much of most plants consists of root. The roots of a rye plant, for example, penetrate the soil to a depth of about 75 cm (30 inches) and may have a total length approaching 650 km (400 miles), with root hairs growing from them whose total length is about 10,500 km (6,500 miles). The root system supports a vast community comprising literally thousands of species of small animals, fungi, and micro-organisms, feeding on the roots themselves, waste products, dead roots and other plant material, and on one another. The soil is a vastly complex ecosystem in its own right – and all its members are made largely from carbon.

Even in the tropics the climate changes with altitude or distance from the sea. This is cloud forest in Costa Rica, on a mountainside where the air is cooler than at sea level and the mist is almost perpetual. Trees are stunted and climbers, epiphytes, ferns, lichens and mosses grow profusely.

Photosynthesis takes place above ground and it uses carbon dioxide. As carbon dioxide is released from the soil it is available for the plants growing above the surface, and much of it is used. What would happen, though, if the plants were cut down and removed at ground level? If they were not replaced by more plants the root systems would decompose as usual, but when they were gone the soil population they supported would begin to starve. They, too, would die and be decomposed until, eventually, the decomposers themselves became short of food. Carbon dioxide would accumulate in the soil, escape from the soil into the air, and in the absence of plants at the surface it would diffuse into the atmosphere.

The tropical forests occupy about 20 million square kilometres (7.7 million square miles). If that whole area, or even a substantial part of it, were to be cleared of vegetation the effect on the climates of the world could be profound and irreversible because of the carbon dioxide that would be released.

What does 'tropical' mean?

The equator divides the northern and southern hemispheres of the Earth, but it is more than a line drawn arbitrarily on a map. Although the geographical and climatic equators do not coincide precisely, the climatic equator marks a belt in which the amount of solar radiation received is at its maximum and, because this belt is presented to the Sun at the same angle at every point in the Earth's orbit, there are no seasons.

The surface is warmed, warms the air above it, and the warm air rises. As it does so denser air from higher latitudes pushes in beneath it, forming a vertical circulation that carries equatorial air away from the equator at high altitude, to descend again in higher latitudes. The rotation of the Earth gives the air a swirling motion as it moves, so forming the trade winds to either side of the equator. Because the equator is the area of lowest atmospheric pressure,

Distribution of tropical rain forests. Tropical rain forests occur in Central and South America, Africa, Southern Asia, and Australia, occasionally spreading just outside the geographical tropics, and within the tropics giving way to other patterns of vegetation in some mountainous or arid areas.

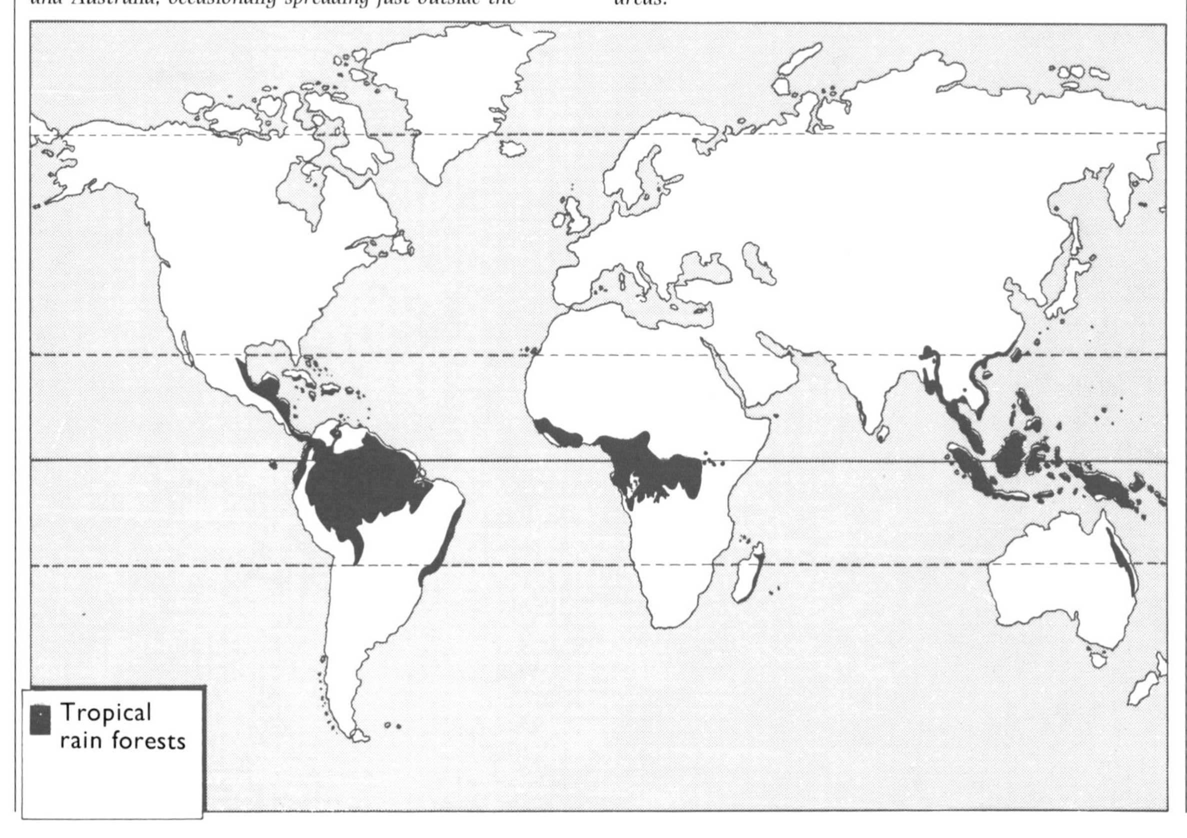

with higher pressure to either side, there is almost no exchange of air across the equator itself, except to some extent in the stratosphere. Air stays inside its own hemisphere.

Because of the high surface temperature, and the fact that most of the equatorial belt lies over the oceans, a great deal of water evaporates. The water vapour rises in the warmed air, the air cools, the water vapour condenses, clouds form, and there is rain. The equatorial belt has a very high rainfall, and the cooled equatorial air that descends again in higher latitudes has lost its moisture and is very dry. The equatorial region is humid, and to either side there is a belt of desert.

The tropics – Cancer in the north and Capricorn in the south – at 23°27′ N and S, define the band in which the Sun appears to be directly overhead at noon on at least one day in the year, but from a climatic point of view the lines themselves mean little. Mountain ranges which affect rainfall locally but also provide habitats for plants at different altitudes, temporary changes in winds, and seasonal changes which become important further from the equator, all produce variations so it is quite wrong to assume that the whole of the tropical belt enjoys the same weather. However, in most of the tropical belt rainfall and temperature are high, and, most important of all, there is no period during the year when it is too cold for plants to grow. The growing season lasts all year.

It also lasts for very many years. At various times in the history of the Earth, and rather frequently in the last few million years, climates have grown cooler and the area covered all year round by ice has increased. All the land in latitudes higher than about 50° N or S has been scoured thoroughly by the ice and all its vegetation destroyed several times. The successive glaciations produced effects that were felt in the tropics, but never to the extent of destroying their vegetation altogether. At no time has the entire face of the planet been covered by ice. Ecologically, therefore, the tropics are very ancient.

Tropical forests

In the southern tip of Madagascar, northern Burma, southern China, north-eastern Australia, and the south-eastern coastal strip of Brazil, tropical rain forest occurs just outside the geographical tropics.

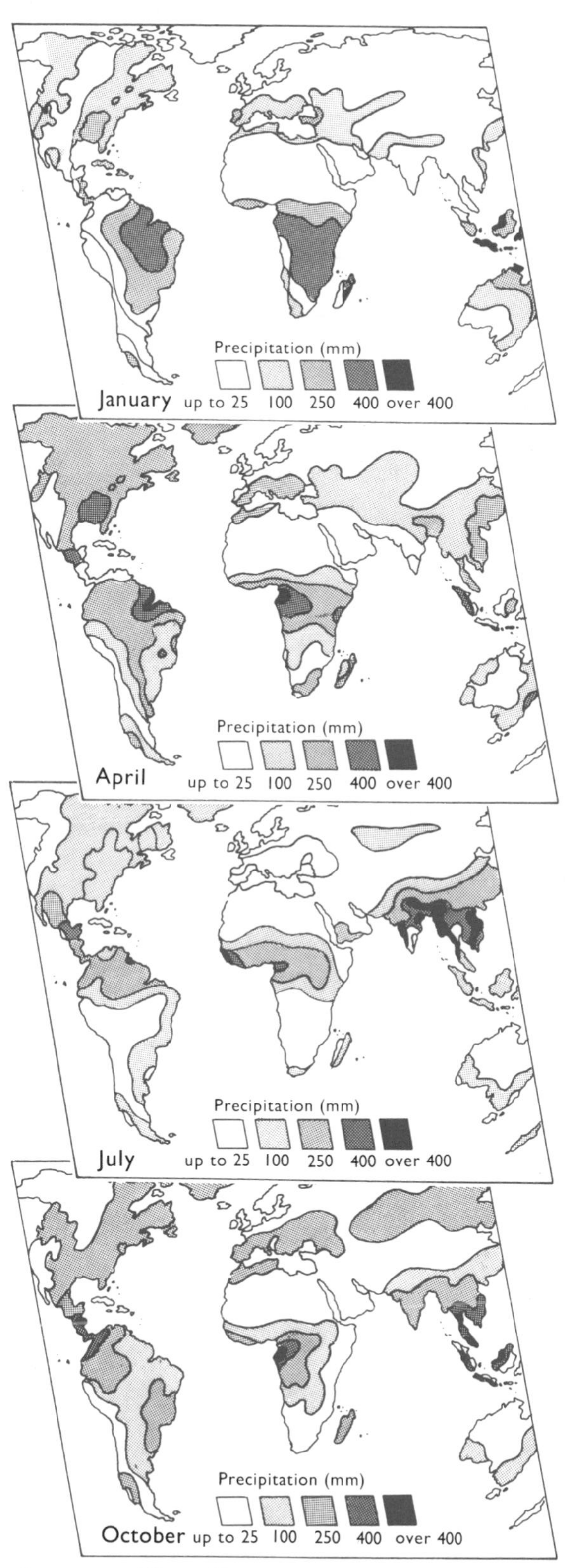

Average rainfall, worldwide, in January, April, July and October.

Apart from these small areas, tropical rain forests occur wholly within the tropics. The largest area is in Central and South America, but the Indo-Malayan rain forest region covers the whole of the Malayan Peninsula, Indonesia, and the surrounding islands. The African rain forest, confined to the southern part of West Africa (Guinea, Ivory Coast and part of Ghana) and the western part of central Africa (part of Nigeria, Cameroon, Central African Republic, Gabon, Congo, Zaire, and part of northern Angola), is much smaller. There is also a strip along the eastern coast of Madagascar.

In these areas, where throughout the year there is warmth and ample water, growing conditions are ideal and the vegetation is luxurious. Although many trees and shrubs are deciduous there is no season during which they all shed their leaves and a particular leaf may survive for several years before being discarded.

There is ample moisture and warmth, but plants need direct sunlight for photosynthesis, and this encourages tall growth and much jockeying for position among the crowns of trees, although it does not pay a tree to protrude high above the canopy, where it will be more exposed to the drying action of the wind. The canopy is complete, so the ground is shaded, and plants growing below the crown are well sheltered from the wind. Pollination by wind is much less important than in forests at higher latitudes. Insects and vertebrates are the important pollinators.

The forest floor is fairly open. The dense, almost impenetrable vegetation through which intrepid adventurers hack their way in movies is found in areas that have been cleared and are being recolonized. Technically, this is 'jungle' and, although its total area may be rather small, it is highly visible to travellers because it is often found where forest has been cleared to make roads. The true forest is arranged vertically. Giant trees project their crowns a little above the main canopy, young trees grow beneath them, waiting for a mature tree to die and fall so they can fill the resulting gap in the canopy, and tree seedlings, shrubs and herbs, up to a metre (three feet) or so tall, grow on the ground, but spaced widely enough to make it easy to move among them. At every level vines and epiphytes – plants that grow on other plants – are common.

The composition of the forest varies according to local circumstances. Along coasts it is often domi-

Top *Inside the Costa Rican cloud forest, shrouded in mist.* ***Above*** *A mangrove swamp in West Java. The trees are all of the pantropical genus* Rhizophora, *which is also the most common tree of the American mangrove swamps.* ***Right*** *Mountain forest in Papua-New Guinea, dominated here by* Nothofagus *(southern 'beech'), growing on limestone on the lower slopes of the mountain.*

Ivory Coast

Brunei

nated by mangroves whose arching aerial roots trap silt and slowly extend the coastline. On mountain sides the species typical of lowland forest give way to those of 'montane' forests, and higher still there is 'cloud forest', where the mist is almost perpetual, temperatures are lower, and trees are festooned in lichens, ferns, liverworts and mosses. The composition also varies geographically. The species dominating the American forest are not those of the Asian or African forests.

Age and diversity

The wide range of habitats, but always with an ideal growing climate, has produced great local diversity in some parts of the tropical rain forests. Every source of nutrient is exploited. Every means to gain access to the sunlight is tried. Yet the boundary between one type of habitat and the next may be impassable, so the species found in one place may be found nowhere else. In Malaysia nearly half of all the native species of plants are found nowhere else in the world, and in the Malayan peninsula, which is about half the area of Great Britain, there are nearly 8,000 species of flowering plants compared with Britain's 1,400 or so.

This diversity of species, for which the tropical rain forests are renowned, should not be exaggerated. Large areas support a dense vegetation but only a very few plant species. Large areas in the Asian swamps, for example, are covered largely by just one tree, *Shorea albida*, and mangrove swamps often contain no more than 25 species of trees and sometimes much less. Not all tropical rain forests are as rich as might be supposed.

Cross section through mature tropical rain forests, in Ivory Coast, Africa and Brunei, Asia. In both forests the tallest trees stand rather more than 40m high, and there is great diversity of trees, as well as of climbers and epiphytes. Note the strong buttress roots to the large African tree (Piptadenia africanum). *The trees around it include* Combretodendron africanum, Dacryodes kleineana, *and* Guarea thompsoni, *with* Strombosia glaucescens, Carapa procera *and others growing beneath. The Asian forest is dominated by dipterocarps, various species of which account for most of the tall trees, although* Ganua palembanica, Dacryodes rostrata *and others grow beside them. In both cases the diversity of plant species (and only trees are shown, not shrubs or herbs) greatly exceeds that found in forests at higher latitudes.*

That said, however, it is also a fact that many tropical rain forests support a vast profusion of species, and it may well be true that there is no plant anywhere in the world that cannot grow in the tropics. In an ordinary forest in temperate latitudes, undisturbed by humans, you might find about ten species of trees in a hectare of land (about four to the acre). In one particularly rich hectare in the rain forest of Brazil botanists once counted 235 tree species (95 to the acre), and in the richest areas half the tree species found in one hectare may be absent from another hectare plot one kilometre away, with different species taking their places.

Animals depend on plants for food and shelter and often build up intimate relationships with particular types of plants. It follows, therefore, that the diversity of plant species is reflected in the diversity of animal species.

The great abundance of species found in much of the tropics is the result of the wide variety of very local habitats and the length of time during which species have been adapting evolutionarily to thrive in them. Parts of the forests have remained essentially undisturbed for the last sixty million years. That has allowed ample time for evolution to proceed at its own pace in relative isolation from the dramatic events taking place in the world beyond the borders of the forests.

The price of antiquity

The forests are ancient, and so are the soils in which they grow. Plant nutrients, supplied from the underlying rocks, are largely gone, washed away over millions of years by rains and rivers to vanish in the seas. Most tropical soils are inherently poor, depleted, infertile.

Most of the nutrients are contained in the living organisms themselves and decomposition is rapid, for the warmth and humidity that favour plants also favour the organisms which decompose organic wastes. The leaves that fall to the ground would take a year to disappear in a temperate-latitude broadleaved forest. In a tropical rain forest they are gone within six weeks. In a temperate forest the soil is rich in plant nutrients because decomposition is much slower, plant growth is slower, and for part of each year plant growth ceases. The clearance of a temperate forest usually provides a very fertile soil for

Above *Slash-and-burn farming makes small temporary clearings, seen here in Guyana. Forest will invade them again if they are left long enough after the farmers leave.* ***Left*** *Trees are felled, useful wood collected, and the remainder fired, clearing the ground and leaving the ash as fertilizer. These farmers are in the Amazonas region of Brazil.* ***Right*** *Tropical rain forest will recover after trees have been taken. Selective cropping has done little harm to this forest, in northern Sumatra.*

agriculture. The clearance of a tropical forest leaves an infertile soil. The luxuriance of the forest is misleading. It does not imply great fertility.

Soils are often from two to 15 metres (six to 50 feet) deep, but have almost none of the fragments of rock from which nutrients might be dissolved slowly. They tend to be rich in clays. They are wet the entire time, so that any soluble nutrient is washed from them quickly, and a short distance below the surface the ground may be waterlogged, which means that organisms needing oxygen cannot survive in it. If the soil is well drained, on the other hand, it is likely to be thin and contain little organic matter, because this will have been washed from the surface.

In this situation plant roots take on a new role. Some trees produce deep tap roots, but most have very superficial roots extending laterally just below the surface. Roots have two purposes, to anchor and support the plant and to obtain nutrients for it. Where plants are sheltered from the wind anchorage is less important, and where nutrients are cycling fast it is essential for each plant to grab what it can while it is there. Many forest trees have roots with very few root hairs – the fine structures through which nutrients enter the plant – and have complex relationships with soil fungi and bacteria which supply them with nutrients direct, perhaps in return for carbohydrates produced by the plants. These relationships are among the many aspects of tropical ecology about which little is known.

The total mass of all the plants taken together is considerable. In Thailand it has been measured at 500 kg to 1,700 kg per square metre (900 lbs to 3,100 lbs per square yard), of which between one-tenth and one-third is accounted for by the weight of plant roots.

Are the forests worth guarding?

The age and ecological richness of tropical forests give them a unique scientific value, but until quite recently they were too remote, too inaccessible to have been studied intensively. They have attracted much more scientific attention in the last half century or so, but vast areas still remain barely explored and mapped, far less studied in detail.

There is much we do not understand about evolutionary processes, about the way organisms are adapted to the circumstances in which they live. This can be studied in temperate regions, but it is easier and more profitable to do so in places where communities have evolved without disturbance over long periods. The tropical forests could furnish us with valuable information concerning the history of life itself.

The forests may also provide us with useful products – apart from timber. While it is true that we are unlikely to miss things we have never had, it is also true that many foods and other vegetable materials we use every day originated in the tropics. The rattan we use in 'cane' furniture is made from the stems of climbing palms found only in the forests of Malaysia. Rubber came from the American tropical forest, and so did the pineapple, passion fruit and guava. Bananas grew originally only in the Asian tropics, as did the mango, and such spices as ginger, cloves, nutmeg and cinnamon that first drew Europeans to that part of the world. Even the farmyard chickens that supply us with eggs and meat came originally from the tropical forests of Asia, and their wild cousins, the Asian jungle fowl, live there still. The list of everyday tropical products is a very long one, and it could be made much longer, for there are many which are collected and used by local people that could be developed for export, and still more have been found that satisfy no local need but could be of great value elsewhere.

How many such products might there be? In 1974 a panel convened in the USA by the National Academy of Sciences met to choose the best plants from a list of 400 that had been nominated by plant scientists all around the world in response to a written enquiry. From the 400 possible plants they chose 36 – almost one in ten – as being plants that could be exploited more widely in the tropics as sources of nutritious food, forage or industrial raw materials. To keep the list short, timber trees and medicinal plants were excluded.

No one has made an inventory and no one knows even approximately how many species of living organisms there may be in the tropics. Without even trying to produce numbers, what seems certain is that an area which may contain nearly half of all the growing timber in the world must represent a large proportion of the world's total pool of genetic information. It is genetic diversity that allows organisms to adapt to changing conditions. The clearance of the forests causes the extinction of unknown species adapted to their particular habitats, and their extinction is a genetic loss. It may matter or it may not, but at least we should know what is there before it disappears.

Is clearance positively harmful?

If large areas of the tropical forests are cleared, the release of carbon dioxide may have serious consequences if it occurs at a time when the atmospheric carbon dioxide concentration is already increasing. The release of carbon dioxide would be offset if the area were resown with trees whose total mass and rate of photosynthesis matched those of the original forest, and it would be partially offset if the land were used for agriculture, but there are risks. Once a tropical soil is cleared it no longer has the protection of the forest canopy, which shades it from direct sunshine and scatters and slows the falling rain. Alternately beaten by torrential rain and baked by fierce sunshine the inherently weak soil is liable to deteriorate rapidly. It can be rendered unsuitable for farming or for forestry in a distressingly short time.

When forests disappear so do their inhabitants. ***Above*** *The South American three-toed sloth* (Bradypus boliviensis). ***Below*** *The Costa Rican slender parrot snake* (Leptophis ahaetulla).

Left Forest that has been cleared and burned in Bahia, Brazil. ***Above*** *Logs being taken from the Bahia forest for milling. This area is the habitat of the golden lion tamarin* (Leontideus rosalia) *one of the world's rarest primates.*

The clearance may distort the cycling of water. Where forests have been cleared high on slopes the rain has washed away the soil which has then polluted rivers and altered the amount of water they carry. In the lowlands, exposing the soil to sun and rain may flood an area, turning it into swamp, or bake it dry, depending on local circumstances. It has been suggested that by reducing the amount of water removed from below ground and released as vapour by plants forest clearance may alter rainfall itself, and so change climates. There is no proof of this, however, and it is unlikely. It has been suggested that the alteration of the colour of the land surface over a large area, as the dark green of the canopy gives way to the paler colours of farm crops or the still paler colour of baked earth, may change the radiation balance of the tropics. More radiation will be reflected back into space, and less absorbed. Again, there is no evidence that such a change has occurred, and it would have little or no global effect unless almost all the forest were cleared.

What is happening?

The tropical forests are being cleared but it is not easy to calculate the rate at which the clearance is taking place, or its precise extent. On the ground appearances can be deceptive. The main road cut through virgin forest causes the destruction of the forest, and makes an appalling mess, but it does not necessarily affect the interior of the forest, which is out of sight of the observer on the road itself, so the area involved is small.

The total area is so large it is difficult to imagine. A modern jet airliner takes several hours to cross the South American tropical forest by the shortest scheduled route.

The best estimates are provided by satellite photographs, which cover large areas and are repeated at regular intervals, so that one photograph can be compared with another. Different types of vegetation can be identified and so changes can be monitored. They show that in the tropics as a whole probably about 0.5 per cent of the forested area is cleared each year. It sounds little enough, but it amounts to around 100,000 square kilometres (nearly 39,000 square miles). That is an area almost the size of Kentucky, or about half the size of Scotland. If the rate of clearance were maintained, the tropics would be cleared entirely of their primeval forests – there would still be plantations – by some time in the next century. That will not happen, because parts of the tropics are mountainous, swampy, and otherwise difficult and, more important, very expensive to work.

The rate of clearance varies greatly from one country to another. In Malaysia, for example, barely half of the original rain forest remains. It has gone to make room for farms, rubber and oil palm plantations, roads and cities, and the trees that once grew there have been used for timber, or burned as fuel. In 1971 it was estimated that the timber from nearly two square kilometres (0.8 square mile) of Malaysian forest was being exported through Singapore every day, and more was leaving through Malaysian ports. In Brazil, on the other hand, the rate of clearance is now well below the world average.

Why?

The tropical forests produce the finest hardwood timber in the world and there is a huge demand for it in the industrial countries of Europe, North America, Japan, Taiwan and Korea, whose own forests grow mainly softwoods. The demand is not for just any tropical hardwood, however, but for particular species. Because of the great diversity of the forests this often means that only one or two trees may be felled in each acre. It sounds as though this should cause little disturbance, but often the result is devastating. Equipment must travel into the forest, and out again dragging the felled trees. This means trees must be removed as tracks are cut for access. Then, as one tree falls it may bring down neighbouring trees. Such selective felling may cause considerable damage. In some parts of the tropics selective felling is giving way to clear-felling, in which an area is cleared of all trees, and all the wood, of all species, is used to make chipboard. This can be less destructive of the forest as a whole, provided it is confined to certain areas.

Land is cleared to make way for farms. This seems natural, for most of the European forests were cleared to provide farmland – and not, as some people suppose, to provide timber for construction or fuel for industry. Although tropical soils are not well suited to agriculture, the climate can be and with care the soils can be managed. The farming may be of little benefit to local people, however. In South America, cattle raised on grassland where forest once grew produce beef a few cents a pound cheaper than beef can be produced from cattle fed on cereals in the United States. To some extent American tropical forests are disappearing to supply the US hamburger market.

Wherever tropical forest is cleared, for whatever reason, poor people are likely to move in and try to start farming. In some countries they are actively encouraged to do so, and in some cases they may be unskilled workers taken to remote regions to participate in a development project, then abandoned, unemployed, when the project is completed. Their farms are usually under-capitalized and consequently poorly equipped, their farming knowledge may be rudimentary, and often they end defeated by sun and rain, hungry, poor, and the depressed inhabitants of what is virtually desert – land from which the forest has been cleared and to which it cannot return.

The reduction in the area of forest increases the pressure on another group of people, the 'slash-and-burn' farmers. This primitive form of agriculture can

be sustained for a very long time under ideal conditions.

An area is cleared by felling the trees, so far as possible making them bring down others as they fall, using such timber as can be used, burning the rest, then planting seeds in the ashes. The ashes enrich the soil with potassium (potash) and for a time good crops are produced, but after a few years tree seedlings start reappearing, weeds proliferate, and yields drop. Then the farmers move to a new area and start all over again. They work in a cycle, visiting one area after another and usually returning to each area after an absence of about twelve years, which allows ample time for tree regeneration and the restoration of the soil. If the area available to the farmers decreases they have no choice but to return to each area more often. This allows the soil too little time to recover after each cropping period and so the system becomes less and less productive until eventually it fails altogether.

Finally, as more and more people choose or are persuaded or even compelled to move into forest areas, forest is cleared simply to supply them with the fuel they need to cook their food and heat their water. The local demand for fuel wood is now a main cause of tropical forest clearance.

What can be done?

In some tropical countries the economic, social and political system is very resistant to change and is dedicated to the preservation of the way of life of a few privileged rich families. They own most of the best land while many would-be farmers have no land, and they control the government, judiciary, and financial institutions. In such countries it is easy enough to recognize the problem, but much more difficult to do anything about it.

More generally, the problem is poverty. Countries

The deforestation of the tropics. The histograms show the estimated area deforested annually in the late 1970s by countries in Central and South America, Africa, and Asia, and those areas as proportions of the total forest area in each country.

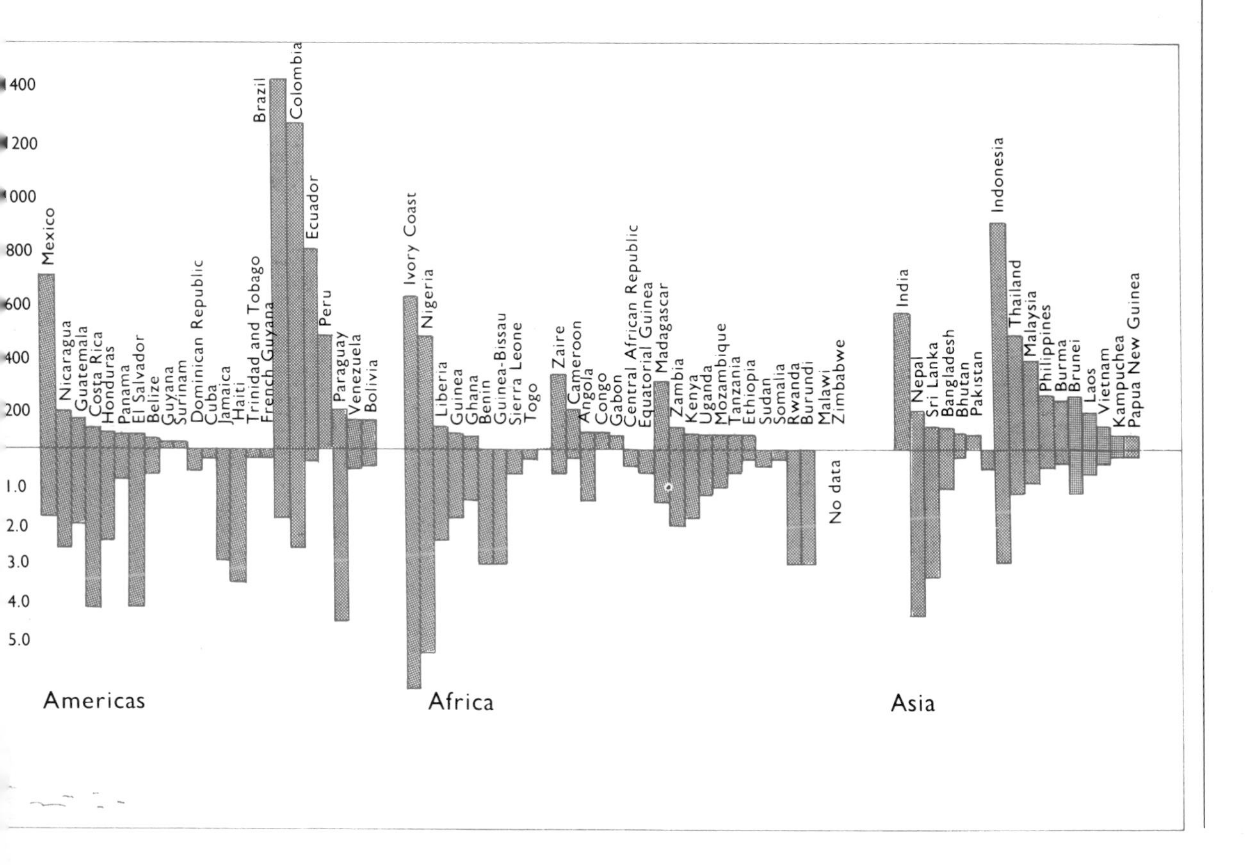

need to export goods to earn the currency with which to buy manufactured goods and set up factories to start industries of their own, or to pay the interest on loans from banks in rich countries. They export their most obvious commodities, taken from their forests, and so exploit their principal natural resource. As often as not, what is taken and how it is taken is dictated by the customer, and the whole operation may be controlled by a foreign corporation, and it is run as cheaply as possible.

It is obvious that if the forests were managed better, not only would large areas of them be conserved, but they would actually become more productive economically. Areas that had been cleared could be devoted to plantations of suitable trees, local factories could process the timber to sell it as finished goods and not simply as logs or sawn planks, and tourism could be encouraged. If poor people could be given simple, cheap, kerosene cooking stoves, one for each family, and a supply of kerosene they could afford, they would need less fuel wood. It would help if they could have even wood-burning stoves that burn wood completely and therefore more efficiently than an open fire. Such stoves are in use in India. It is obvious, and there is no shortage of helpful suggestions, but suggestions and exhortation are not enough.

If there is to be change, if the tropical forests are to be protected, the reforms must be made first within the rich countries. It is there that patterns of international trade are determined and tropical development projects approved or disapproved, financed or rejected. We cannot stop importing tropical produce, for that would make the poor still poorer, but we could agree to pay much higher prices, as a kind of rationing of output that does not reduce the income of the producer. We could provide much more help, on much more favourable terms, for the development of alternatives to deforestation. We could exert greater control over the operations of corporations working out of our own countries. The US authorities could tax beef imports from tropical forest areas – and benefit American farmers.

There is much we could do, and at last there are signs that we may be starting to do it. The risk facing the tropical forests has been noted, and politicians have begun to listen to their scientists. The United Nations Food and Agriculture Organization is preparing an Action Programme to protect tropical forests. Investors in tropical industries are being advised by the Washington-based World Resources Institute on the desirability of forest conservation. An International Timber Agreement will come into force in the near future. It is the first international commodity agreement to include a clause on conservation and it may help to improve tropical forest management.

Some of the tropical forest area has disappeared already and it is inevitable that more will go before the clearance is checked, but it can be checked if that is what we want. We can make it possible for much of the forest to remain, to be studied, visited, and enjoyed by our children and their children, not as some kind of museum, but as a dynamic, developing, evolving system. We have to help impoverished people in tropical countries to earn a decent living without being forced to sell cheaply the resources on which they depend. Manage resources well and they will last for millions of years. Dispose of them and the gain will be small and when it is gone the poverty will be real.

If this is a cause that appeals to you – and most scientists believe it is important – you might try to bring closer together the organizations devoted to conservation and those devoted to economic development.

*Undamaged forest in Ecuador (**right**), dominated by many species of tall trees whose foliage forms a complete canopy, creating a different climate for the plants beneath. The forest environment has also provided humans with a satisfactory habitation for centuries (**above**).*

CHAPTER 7

Acid rain

As the Sun warms the surface of the Earth water evaporates from rivers and lakes, but most of all from the seas and oceans. Green plants draw water from beneath the soil surface, carry it upward to their leaves, and it evaporates from their open stomata, or 'pores'. As it evaporates from the leaf stomata the water is replaced from below, so there is a constant flow of water through plants from the ground and into the air. Over the planet as a whole, every day more than 1,000 cubic kilometres of water evaporate from the surface.

The water vapour rises in the warm air, is moved as the mass of air containing it moves, and as it rises it cools, eventually condensing to form clouds. Every day as much water falls on the surface of the planet as evaporates from it. This is the 'hydrological cycle'.

The air into which the water evaporates and in which it is carried consists mainly of nitrogen and oxygen, but there are other substances as well, present in very small amounts. There is a little carbon dioxide. There are particles of soil carried aloft by the winds blowing over the dry regions of the world, and particles of dust and soot. There are oxides of nitrogen, some released by volcanic eruptions and some formed by the oxidation of nitrogen in the atmosphere itself. This reaction requires large amounts of energy, and the energy is provided by the thousands of lightning flashes that occur every day. There are oxides of sulphur, also released by volcanoes, and there are tiny crystals of salt, blown into the air as sea spray and left in the air after the water that carried them has evaporated. All of these substances can and do mix with the atmospheric water vapour and are carried by or dissolved in the water droplets in clouds, to be washed back to the surface in rain, hail or snow.

When oxides of nitrogen react with water they form acids, including nitric acid. When oxides of

The Schwanwald, in the West German Black Forest. Vast areas of coniferous forest in central Europe are suffering like this but what causes the damage?

sulphur react with water they, too, form acids, including sulphuric acid. If reactions with other atmospheric constituents separate the atoms of sodium and chlorine which form common salt, and the chlorine reacts with water, it forms acids, including hydrochloric acid. Since the natural constituents of air are present all the time, although the concentration of those released by volcanoes varies according to the amount of volcanic activity, rain and snow themselves are usually rather acidic.

Chemical acidity is measured on the pH scale. 'pH' stands for 'potential of hydrogen' and the scale measures the concentration of hydrogen ions in a substance, in turn as a measure of the activity of the hydrogen. On the pH scale pure water has a value of 7, and is neutral. A pH of less than 7 indicates acidity, a pH of more than 7 indicates alkalinity. Ordinary rain and snow have an average pH of about 5. Thus any discussion of 'acid rain' must start from the fact that rain tends to be acid naturally.

The start of industrial air pollution

In 1852, R. A. Smith presented a paper 'On the air and rain of Manchester' to the Manchester Literary and Philosophical Society. In it he reported that rain falling in the Manchester area contained sulphuric acid, the amount of acid increasing the closer one approached to the city. Probably this was the first written account of the acid rain phenomenon, but equally probably the phenomenon was not new. It must have existed at least since the beginning of the nineteenth century. Estimates made in 1913 suggested that rain contained relatively large amounts of sulphuric acid over much of industrial northern England.

Over how large an area was unusually acid rain falling? No one really knew, although the agricultural research station at Rothamsted, in Hertford-

Below *Small particles are seldom airborne for long, but they can travel considerable distances. Winds are blowing at up to 160 km/h in this Colorado dust storm. Some of the dust may be carried hundreds of miles in a day or two.*

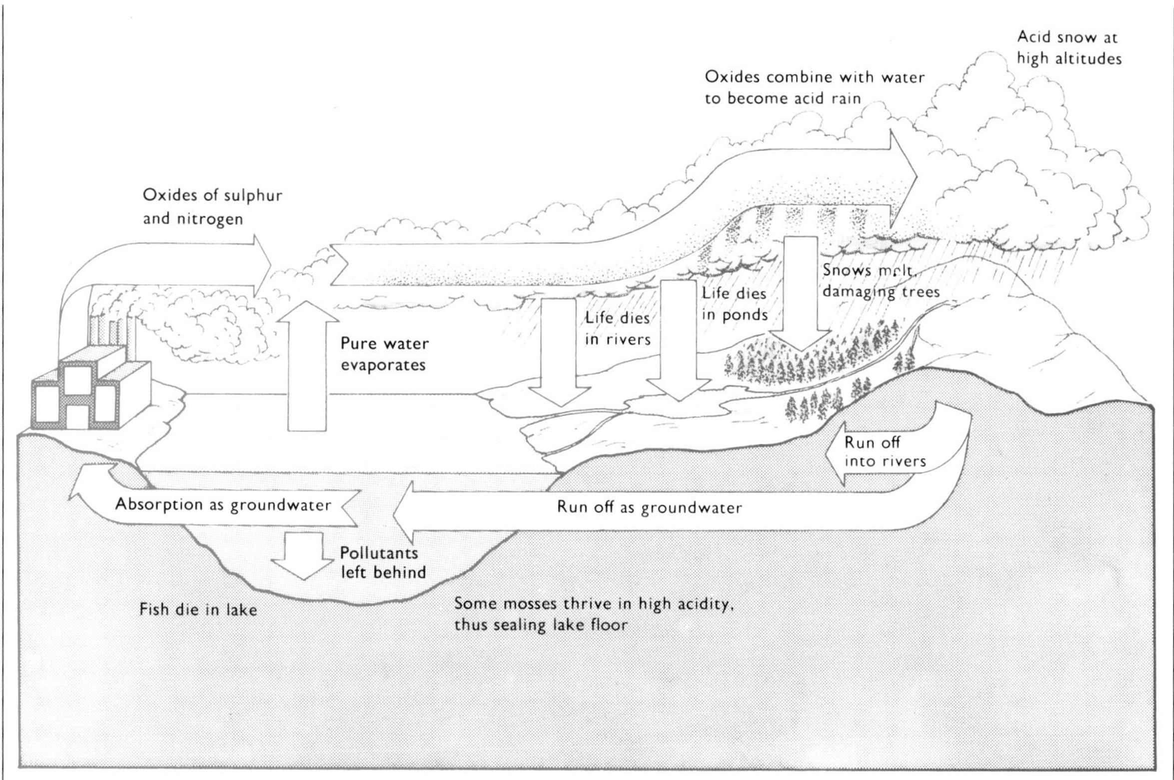

Movement of atmospheric pollutants

shire, had been analysing rain since 1853 and it seemed that rainfall in southern England was not being affected. It was not until the 1950s that systematic monitoring began over northern and western Europe, and not until the early 1970s that rain with an abnormally low pH was observed over the north-eastern United States and parts of south-eastern Canada. Today most atmospheric scientists regard acid rain as a global problem. It has been reported from China and the USSR, but it is especially marked in Scandinavia, central Europe, and parts of eastern North America.

How do the pollutants travel?

Pollutants enter the air, but once airborne they are subject to many processes that make it difficult to say with certainty what will happen to them. They may be carried upward in warm air and then carried in bulk by prevailing winds. They may diffuse through the air around them, so the air dilutes them but spreads them as well. They may react chemically with one another or with substances in the air already, so their composition changes.

The reactions are complicated and there are many of them. There are several ways in which sulphur dioxide (SO_2) may be converted to sulphuric acid (H_2SO_4), for example, and the conversion of sulphur dioxide to sulphuric acid is simple compared with the processes by which nitrogen oxides (known generally as NO_X) are changed to nitric acid (HNO_3), because many of the nitrogen reactions keep reversing, so oxygen is being added to and removed from the nitrogen repeatedly until eventually nitrates (NO_3) are formed.

Air pollutants have a short life. They may come into direct contact with solid objects to which they adhere. This is called dry deposition. Cloud or fog may be carried along at ground level, and as it wets the solid surfaces in its path the dissolved or water-borne pollutants are deposited on them. This is one form of wet deposition. The other kind of wet deposition occurs when pollutants are washed to the surface in rain or snow. In the strict sense only this is acid rain, but in recent years the use of the term to cover all three kinds of deposition has confused the issue.

Depending on the amount of precipitation (the fall of rain, hail or snow) that occurs while they are being transported, pollutants usually remain airborne for three to five days. The determination of their origin after they have landed is based on studies of the recent movement of air masses and on tracing them backwards for up to five days.

How much pollution is there?

Over the world as a whole it is estimated that about 60 per cent of all the sulphur dioxide in the air is released naturally. Humans add the other 40 per cent, amounting to about 100 million tonnes of sulphur a year. The amount of nitrogen oxides released naturally is known with much less certainty. Scientists accept that it is significant, but estimates of the ratio of natural nitrogen oxides to those released by human activity range from 15:1 to 1:1. Oxidation with the energy of lightning is believed to produce about one-fifth of all atmospheric nitrogen oxides.

Volcanoes and lightning flashes are distributed widely over the surface of the planet, so although the total contribution may be large, in most cases it occurs as fairly small amounts in any one place. Industrial emissions, on the other hand, are highly concentrated, so locally their contribution may be very much greater than it is globally.

What we do know, with rather more certainty, is the amount of sulphur dioxide and nitrogen oxides we release ourselves and the way those releases have changed over the years. In the United Kingdom, for example, 2.8 million tonnes of SO_2 were released in 1900, and the amount increased steadily until it peaked in 1970 at 6 million tonnes. After that it started to fall, a little hesitantly at times, until in 1982 only 4 million tonnes were released, less than had been released at any time since the mid-1940s. More than half of it came from coal-fired power stations. Nitrogen oxide emissions rose steadily until the early 1970s and have remained fairly constant since then, at around 1.7 to 1.8 million tonnes a

Emission of sulphur dioxide in Europe and the western USSR

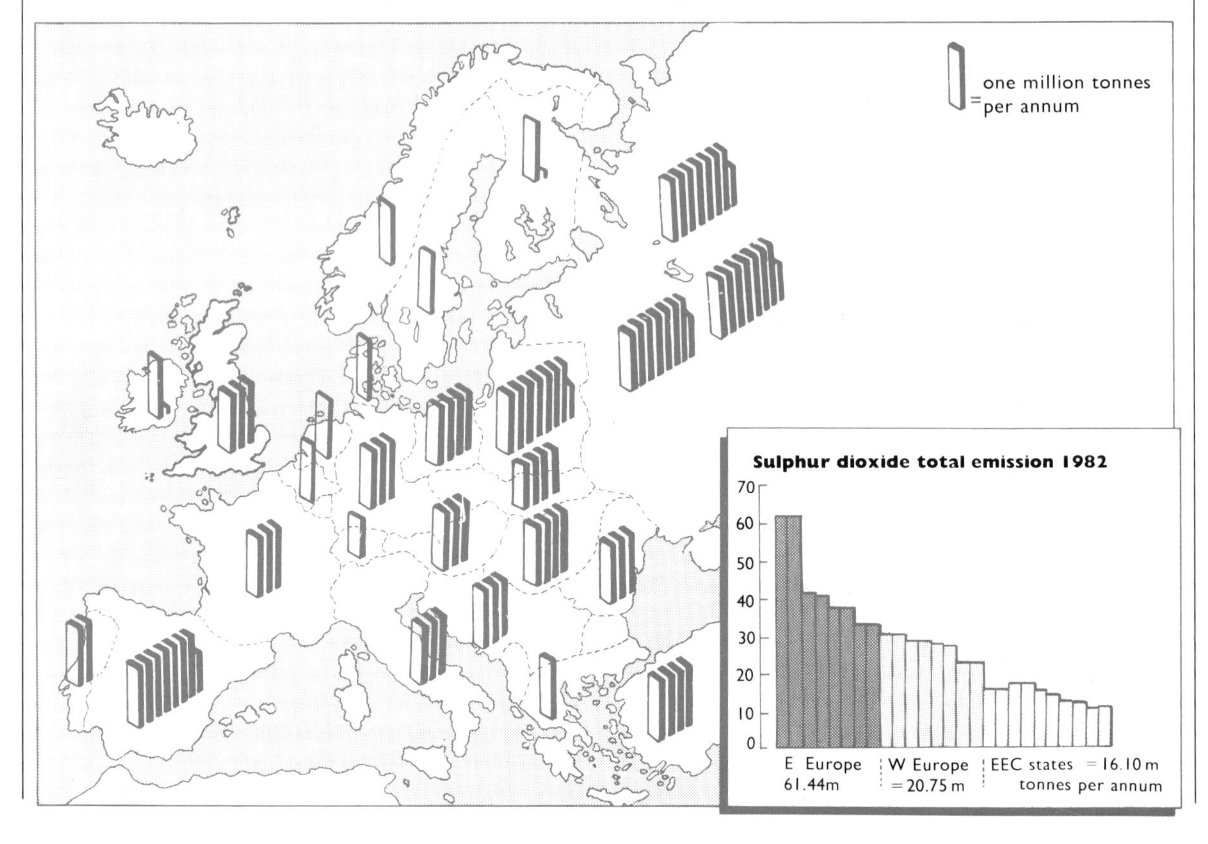

year. The pattern in the United States was similar. Sulphur dioxide emissions reached their peak, of around 32 million tonnes, in 1970, nitrogen oxides peaked in 1973 with the emission of about 25 million tonnes, and since then sulphur dioxide emissions have decreased by 7.6 per cent and those of nitrogen oxides by 10.6 per cent. The largest European source of both sulphur dioxide and nitrogen oxides is the western USSR (16 million and 5 million tonnes respectively in 1978). Czechoslovakia, both Germanys, France, Italy, Poland and the United Kingdom altogether released nearly 27 million tonnes in the same year. In the case of nitrogen oxides, West Germany released 3.4 million tonnes in 1978. Apart from the USSR that was more than twice as much as any other European country.

Does it matter?

The fact that strong acids are being formed in the air does not necessarily mean those acids cause damage. Both sulphur and nitrogen are essential nutrients for plants, for example. Twenty years ago agricultural chemists were pointing out with some glee that because of industrial pollution by sulphur it was possible to reduce the amount of sulphur in fertilizers. The pollution was believed to be feeding farm crops. It is possible, therefore, for air pollution to be beneficial, and if the pollutants are utilized by plants they are also removed from the environment and made harmless to other organisms which might be sensitive to them.

Today this would be a minority opinion, not to say a somewhat eccentric one. Most scientists are agreed that strong acids are at least potentially harmful. It still does not follow that they will cause actual harm. If they are at low concentration they may have little or no adverse effect and if they fall on the land or on water rather than on plants they may be neutralized by alkaline soil or water. Soils, rivers and lakes with a pH of 7 or slightly higher are widespread in areas underlain by limestone and chalk.

Where the pollutants fall directly on plants their effect depends very much on the species of plant. Although sulphur and nitrogen are nutrients, an

Deposition of sulphur in Europe and the western USSR

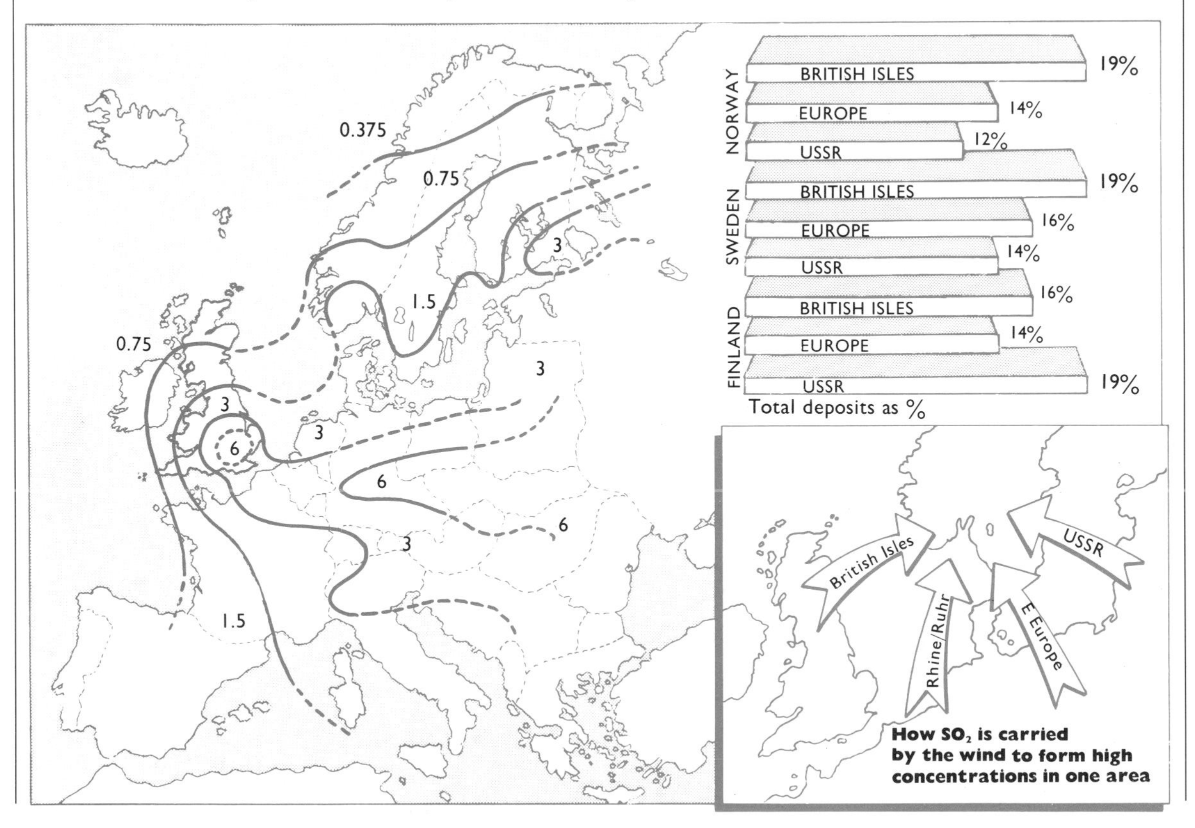

How SO_2 is carried by the wind to form high concentrations in one area

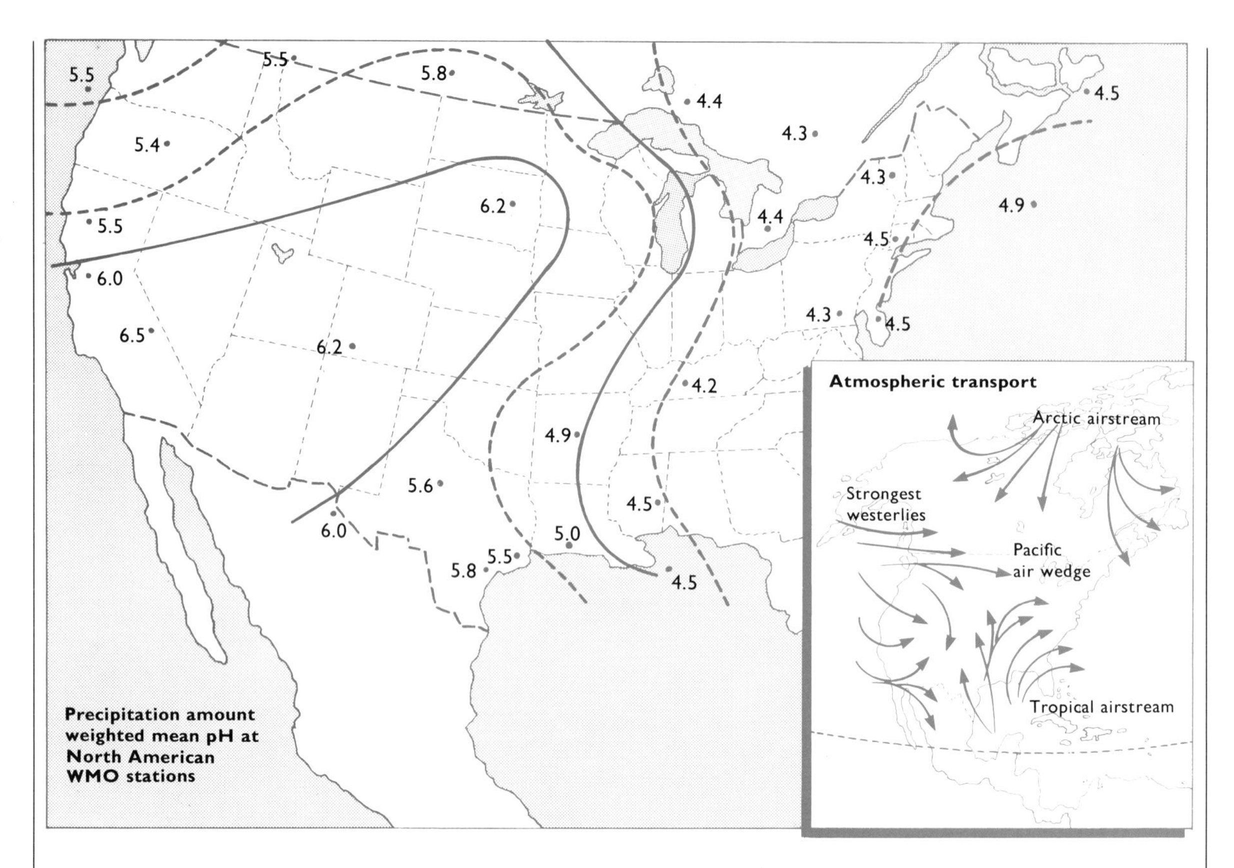

Rainfall is usually slightly acid because of carbon dioxide and other naturally occurring acids dissolved from the air by water droplets in clouds. The burning of fuels adds more acids, as sulphur dioxide and oxides of nitrogen. These are carried in the air and take part in complex chemical reactions which may lead to them being dissolved in water droplets to fall with rain or snow, or to their direct deposition on surfaces from dry air. Because the length of time pollutants usually remain airborne is known, a map of the acidity of rainfall, based on information collected during a short period, can be used in conjunction with one showing prevailing winds during the same period to trace the source of the pollution.

excess of them can be disruptive. In some species sulphur dioxide may damage cells on the surface of leaves and cause stomata to open. The pollutant enters through the stomata, and dissolves in water surrounding cells just below the surface. This causes water to flow out of the cells under osmotic pressure to equalize the strengths of the solutions inside and outside the cells, which makes the cells less rigid (because they now hold less fluid) and so causes the stomata to open still further. In other species, at higher concentrations, or under drier conditions, sulphur dioxide can cause stomata to close. Nitrogen dioxide may have similar effects, although these have been studied less. Any interference with the opening and closing of stomata is likely to injure a plant by disrupting the way water moves through it, by reducing the amount of carbon dioxide that can enter for photosynthesis, or both.

Plants allocate the material they manufacture to the growth of shoots and roots. In plants that have absorbed sulphur or nitrogen dioxide more material is used for shoot growth than for root growth. This leads to an increase in leaf area, but a smaller root system. The larger leaf area increases the transport of water from the soil, but the smaller root system collects it from a smaller area, so very locally the soil can be dried out so the plant is no longer able to obtain as much water as it needs. Sulphur dioxide

Right *Effects of increasing acidity (falling pH) on freshwater organisms. The diagram is approximate, because local variations in conditions may increase the ability of some species to survive. Pike can sometimes live at pH 4, for example.*

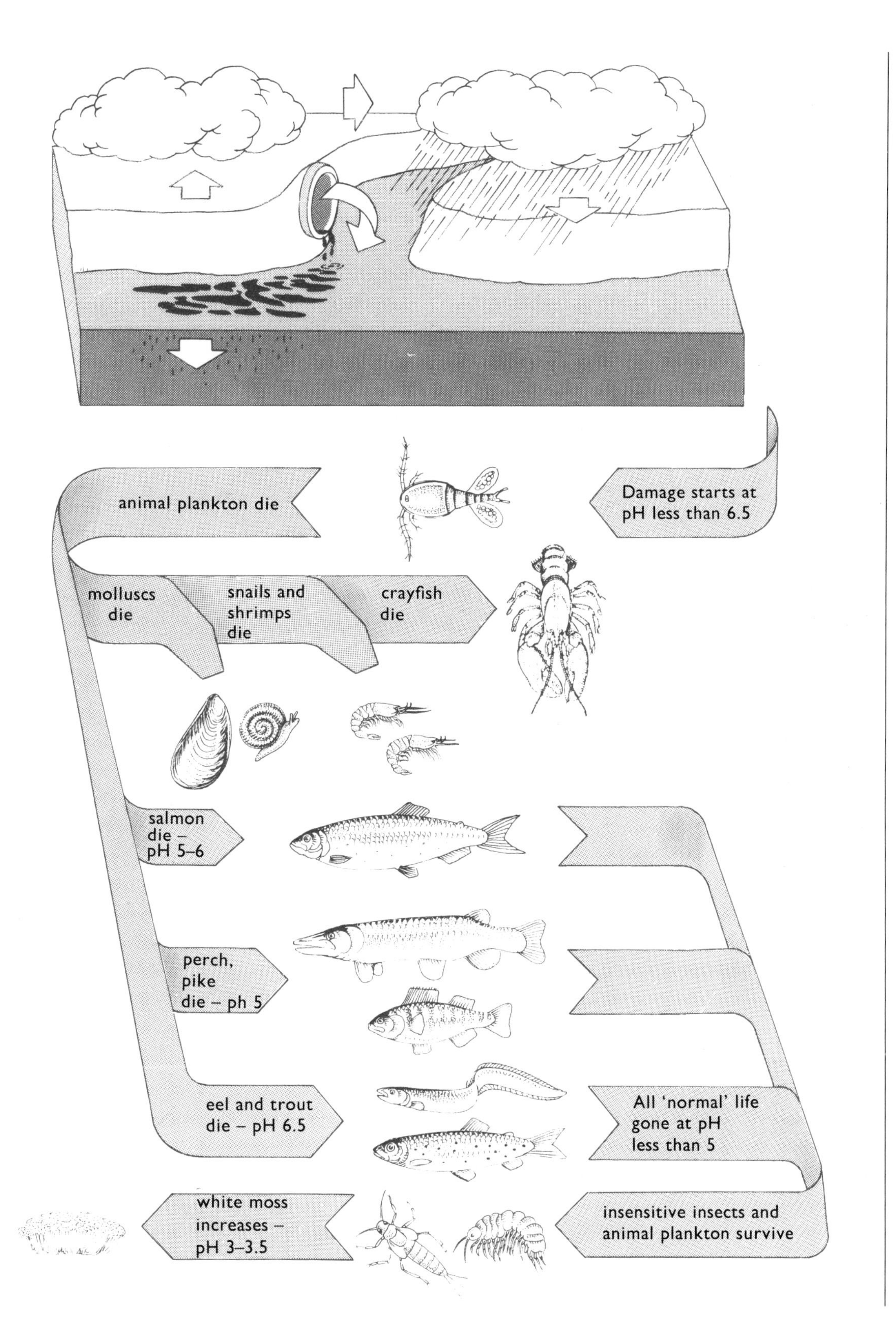
Damage starts at pH less than 6.5
animal plankton die
molluscs die
snails and shrimps die
crayfish die
salmon die – pH 5–6
perch, pike die – ph 5
eel and trout die – pH 6.5
All 'normal' life gone at pH less than 5
insensitive insects and animal plankton survive
white moss increases – pH 3–3.5

also affects the ability of a plant to withstand frost. Frost damage is more likely to occur, and to be more severe, if the plant has taken in sulphur dioxide.

Acid deposition may also affect freshwater lakes, but here the chemistry is even more complex. Acid lakes, with a pH of 5 or less, are fairly common, especially in the uplands and where soils are rich in peat. Near coasts sea water may intrude, altering the salinity of the fresh water and temporarily making it more acid. Changes in land use, such as the ploughing and fertilizing of previously uncultivated land, but especially the great increase in the area of forest in Britain in recent years, make nearby rivers more acid, and lakes are fed by rivers.

The presence of certain substances in a solution, in freshwater lakes, the most important of which is bicarbonate, will make the solution resistant to changes in pH when an acid is added to it. This is called 'buffering'. The effect on a lake of acid deposition therefore depends on how well the water is buffered.

Some are poorly buffered, their pH falls, and the effect on aquatic organisms is dramatic. Many of the small or microscopic species disappear. If the pH falls below 4.5 trout cease to produce the enzyme responsible for breaking down the outer coating of their eggs, trapping the larvae, and so preventing reproduction. Thus, as older trout die they are not replaced and in time the trout disappear. Salmon are also highly sensitive to reduced pH. At pH 6 some species of trout suffer. At pH 5–6 rainbow trout disappear, and other relatives of the salmon and trout suffer. At pH 4.5–5 carp suffer, at pH 4–4.5 a long list of fish species are being harmed, although pike can still breed. At pH 3.5–4 most fish, but not pike, are gone, and at pH 3–3.5 only a few invertebrates and some plants survive. Alkalinity can be just as harmful. Fish suffer at a pH higher than 9 and none can survive a pH higher than 11.5.

In more acid conditions aluminium forms compounds that enter water. They irritate the gills of fish, impairing respiration, and the results can be fatal. Aluminium may replace phosphorus – an essential plant nutrient – and so kill certain plants, but at the same time aluminium causes the precipitation to the lake bottom of humates, which are complex salts formed by reactions involving humic acids, in turn the products of the decomposition of organic matter. Their precipitation makes the water clearer, so light penetrates to a greater depth, and this allows more photosynthesis among the phytoplankton – microscopically small single-celled plants. So although many species are killed, the total amount of living material in the lake may stay the same.

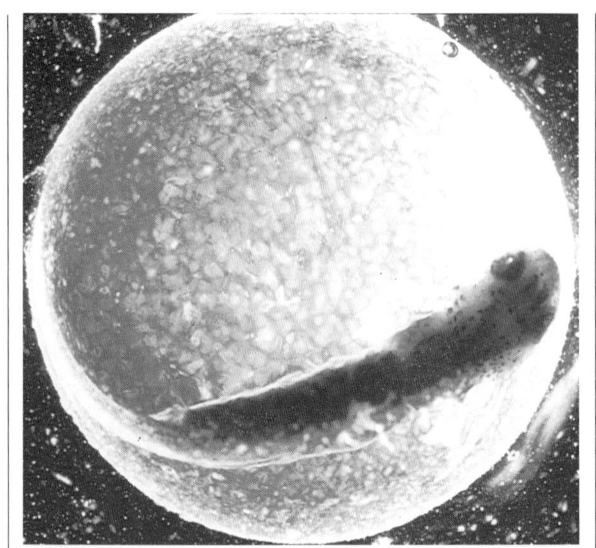

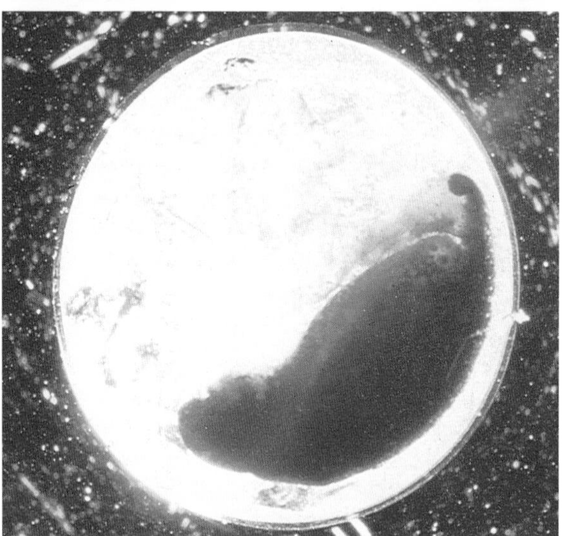

American mole salamanders live in water until they are adult. The Ambystoma maculatum *embryo* ***(top)*** *is in ordinary water. The other embryo* ***(above)*** *is deformed. It is in water matching that in acid rain areas.*

Is there a problem?

In the 1960s Swedish scientists became concerned at the high sulphur concentrations in central Stock-

Airborne acid has damaged stonework in every industrial city, but emissions of sulphur dioxide and nitrogen oxides have fallen and restoration work can begin, as on the cloisters of Canterbury Cathedral, seen here. Can such emissions be damaging trees?

holm. These were damaging buildings by corrosion, as similar pollution has damaged buildings in most of the industrial cities of Europe.

At about the same time evidence was accumulating of the increasing acidification of lakes and ground water, especially in southern and central parts of their country, with pH levels falling below 5. In much of Sweden the underlying rock provides little chemical buffering, so Swedish waters have little protection. It is estimated that about one-tenth of the entire lake area of Sweden has been affected and freshwater fish populations have been seriously reduced. In 1968 the Swedish government took steps to reduce sulphur emissions from oil-fired heating installations and since then have reduced emissions further. Nitrogen oxide emissions were regulated in the 1970s, with the introduction of emission limits for new cars.

Norway and Finland were suffering similar problems and since the early 1980s these three Scandinavian countries have collaborated in urging general reductions in emissions, especially of sulphur, throughout Europe. They had calculated that no matter what steps they might take themselves, much of the pollution affecting them was carried by moving air from other countries, and in particular from the United Kingdom. Although three-quarters of the sulphur dioxide deposited over the United Kingdom was produced in the United Kingdom, in Norway the situation is reversed and the United Kingdom is the largest single source of sulphur dioxide. Acidification became an international issue.

In the early 1970s effects were observed on forests in West Germany. Today about half the total forest area may be affected to a greater or lesser extent. In a country where forests are loved, feature largely in

Left *Damage attributed to acid rain begins on the lower branches and moves upward and outward, until trees are left with a 'stork's nest' crown.*

Above *The tree is dying back, as its needles start to go yellow. Soon they will fall, denuding the branches. The tree may not recover.*

folklore, and are central to concepts of a beautiful countryside, their decline is now by far the most important environmental issue in the country.

The German forest decline, probably typical of what has happened in Czechoslovakia and other parts of central Europe, began with just one species, silver fir (*Abies alba*), which was seen to be dying back. This was neither important nor unusual. Silver fir accounted for only two per cent of the forest area of West Germany and it was known to die back in this way from time to time. However, the problem increased, most dramatically in 1976, and now virtually all stands of this species are affected. The needles become discoloured and then fall, starting low on the tree and then moving higher, and from inner to outer branches, until the crown of the tree stands on top of a bare stem, like a stork's nest – which is the name given to the phenomenon. After that the lower part of the crown dies, at the base of the trunk the wet heartwood starts to spread into the sapwood, the roots are damaged, and little by little the entire tree dies. By the late 1970s trees other than silver fir were being affected. Scots pine (*Pinus sylvestris*), Norway spruce (*Picea abies*), and then broad-leaved species such as beech (*Fagus sylvatica*) began to suffer. Together these species account for more than three quarters of the total forest area in West Germany, so if the present trend continues the problem will become very much more serious even than it is now, although there is a tendency for older trees to be injured more severely than younger ones, and for trees at high altitudes to suffer more than those at lower levels.

Lakes and forests are also being affected in Canada, and the Canadians blame pollutants carried across their border from the industrial northern states of the USA. Like the Scandinavians, they demand political action in another country for the protection of their internal environment.

Is the diagnosis correct?

When the issue of acidification first emerged it was assumed that the principal cause was sulphur dioxide, a traditional atmospheric pollutant. Most of the research was devoted to the production, emission, long-range transport, and biological effects of this compound. Then doubts began to appear.

Many species of lichens are extremely sensitive to sulphur, and die while concentrations are still quite low. This fact has been known for so long and is so well-established that most investigations of sulphur pollution begin with a search for sulphur-sensitive lichens. In Britain such lichens vanished from most industrial areas during the nineteenth century. Today, when the air is cleaner, they are returning.

Studies of the German forests found no shortage of sulphur-sensitive lichens. When the air in the forests was analysed the initial impression was confirmed. The sulphur dioxide concentration was low – much too low to injure trees – but levels of ozone (O_3) were abnormally high.

Even if sulphur dioxide were responsible for the damage, no one could explain satisfactorily why the damage should occur at a time when sulphur dioxide emissions were falling, and had been falling for several years. Professor B. Ulrich suggested in 1983 that there might be an initial phase during which pollutants accumulated in an entire ecosystem, and a second phase during which effects of those pollutants could be observed, and the two phases might or might not occur at the same time. If he is right, sulphur could have accumulated in the central European forests over a number of years. At first it would have stimulated growth, but then the increasing acidity of the soil would have started damaging roots and soil micro-organisms. Then the accumulation of pollutants in the bark and leaves of the trees would reach toxic levels, and in the final phase trees would have died back, producing obvious, but by this time irreversible symptoms. At present no one knows whether this hypothesis is correct.

Ozone concentrations in the forests were high, similarly high ozone levels have been found in parts of the United States, and ozone is known to damage trees, although the mechanism by which it does so remains uncertain. Ozone is highly reactive and also plays an important part in the oxidation of sulphur and nitrogen dioxide to sulphates and nitrates respectively. These may then change further into sulphuric and nitric acids. In the lower atmosphere ozone is produced mainly by the action of strong sunlight on unburned hydrocarbons emitted largely by vehicle engines – which are also an important source of nitrogen oxides.

The studies relate to central Europe. No conclusive evidence has been found to show that damage to Scandinavian forests is caused by acid deposition – although damage to lakes is.

There are other possibilities, too. 1976, the year in which tree damage in West Germany was so pronounced, was a drought year. Were the trees suffering from lack of water? Were they suffering from pests or diseases? Were the trees being grown badly, or in the wrong places? Such explanations as these are inadequate to account for the whole of the observed effect, but they may have contributed, now and then, here and there.

We are left with many uncertainties, and the scientists most involved with the search for answers urge caution, and more research.

Might the cure be worse than the illness?

There is strong international pressure for all countries to reduce their emissions of sulphur dioxide by 30 per cent by the end of the century. Most scientists agree that this would reduce acid deposition – but by how much?

It is now known that much of the sulphur is released not by industrial emissions but by microscopic marine organisms, as dimethyl sulphide that is oxidized to sulphur dioxide particles on to which water vapour condenses. This is the principal mechanism by which clouds form over the oceans. Industrial pollution is a factor, but in Scandinavia, for example, most of the pollution occurs locally, with some contribution from eastern Europe and only a minor contribution from Britain, long regarded as the villain in the case. To reduce by one tonne the amount of sulphur reaching Scandinavia from Britain, British emissions would have to be cut by 53 tonnes for Sweden and 46 tonnes for Norway.

Sulphur dioxide is released industrially by the burning of coal containing sulphur. The sulphur can be removed, but there are costs. Coal can be washed, but this can mean up to ten tonnes of coal are lost for every tonne of sulphur removed. Or the coal can be burned and sulphur dioxide extracted from the flue gases. Most methods for achieving 'flue gas desulphurization' involve passing the gases through water containing pulverized limestone. The limestone is calcium carbonate ($CaCO_3$), which reacts

Lakes that are poorly buffered can be improved by the addition of lime (calcium hydroxide) which neutralizes some of the acid. The helicopter is spraying lime over a Swedish lake. Opinions vary about the effectiveness and practicality of such treatment.

with sulphur dioxide to produce carbon dioxide (CO_2) and calcium sulphate ($CaSO_4$) – gypsum.

The limestone itself must be quarried. In the United Kingdom the most likely source is Derbyshire – in the Peak District National Park – and large amounts may be needed. One power station, burning four million tonnes of coal a year, might use half a million or more tonnes of limestone a year. The impact on the environment in the quarrying area would be huge.

The cost of removing sulphur from fuel or from flue gases would be passed on to electricity consumers, in the United Kingdom probably amounting to an increase of less than four per cent for domestic consumers and rather more for industrial consumers. The cost would be bearable, but it might alter the economics of power generation. Already nuclear generation is generally held to be cheaper than coal-fired generation. Any increase in the cost of coal-fired generation would tip the balance even more strongly in favour of nuclear power.

The demand for a reduction in sulphur dioxide emissions is accompanied by demands for a reduction in emissions of nitrogen oxides, mostly from cars. This is likely to have a much larger beneficial effect, especially in West Germany, and although there is disagreement about the best way to achieve a significant reduction the desirability of it is accepted.

The acid rain controversy illustrates the complexity of many environmental problems. The damage is on a large scale and sensational. The cause is popularized as being quite simple, and responsive to well-known and tested treatments. This popular conception generates pressure for immediate political action.

Yet in truth the cause is subtle and not at all well understood. Indeed it is still so poorly understood that it is difficult even to estimate the extent of the damage – some of which, but we cannot tell how much, may not be due to acid deposition at all. The proposed remedies may be less efficacious than the campaigners suppose, but certainly they will be expensive and will cause much environmental disruption of their own. Environmentalists will protest strongly at any large increase in limestone quarrying in a national park, and past experience suggests they will protest no less strongly at the idea of changing from coal-fired to nuclear power generation, although this is the conclusion to which the logic of their argument must drive them.

Two lessons we might learn from all this are distinctly ecological. The first is that the natural world is a great deal more complicated than it seems. The second is that everything has its price: if you desire an environmental gain in one place you should be prepared to suffer an environmental loss somewhere else. The third applies to scientific issues more generally: simple solutions composed for complex problems are nearly always wrong.

CHAPTER 8

Nuclear power and the alternatives

If we are to rely less on the burning of carbon-based fuels to supply us with energy we will have to find an alternative, and the alternative about which we hear most is nuclear power. It is also the alternative about which most popular concern is expressed. As the nuclear power battle – for it cannot truly be called a debate – has proceeded the facts have been obscured by propaganda. It is not surprising, therefore, that many ordinary people are frightened by a technology they do not understand, which is represented as being in some way 'unnatural', and which is said to pose quite unique threats.

The solar system formed when a cloud of matter condensed some billions of years ago. It started to condense because it was perturbed and the perturbation was probably due to a collision between two clouds. A cloud of cold matter and a cloud of relatively hot, energetic matter collided. The hot cloud was produced by a 'supernova event', in which a large star exhausts its fuel, starts to collapse, and then casts off its outer layers in a tremendous thermonuclear explosion. We are sure this is what happened because it is only at the temperatures and pressures of a supernova event that elements heavier than iron can form, and the solar system is rich in elements heavier than iron.

The cloud from which the solar system formed was the product of a thermonuclear reaction, and the light and heat we receive from our own Sun is also produced by a thermonuclear reaction, although a less fearsome one. We and our world are made partly from the products of nuclear reactions,

The Dounreay fast reactor site. At present Britain has only one fast breeder reactor, a prototype on the coast at Dounreay, Caithness. Since 1975, it has been providing design information for a planned full-sized 1200 MW commercial demonstration fast reactor.

and we are warmed every day by a vast thermonuclear reactor.

The interior of the Earth is hot. Its heat is produced mainly by the decay of radioactive substances. Our own planet is therefore a giant nuclear reactor. If the nuclear power industry produces electricity by using the heat released through the controlled decay of radioactive substances to generate steam that drives turbines, clearly there is nothing unnatural about it. Nor is there anything unnatural in the idea of building thermonuclear reactors, which exploit a process much like the one taking place in the Sun. Indeed, you could argue that nuclear power is more 'natural' than the burning of fossil fuels. Radioactive decay occurs constantly; it is rare for fossil fuels to burn spontaneously.

Fission and fusion

There are two processes by which energy can be obtained from nuclear reactions: fission and fusion. The fusion, or thermonuclear reaction is the one that mimics the Sun. In it, the nuclei of very light atoms, such as deuterium and tritium (two forms of hydrogen) are made to fuse together to form heavier helium nuclei, or deuterium nuclei themselves are made to fuse to form tritium, or lithium is made to fuse with deuterium to form helium. In each case the fusion is accompanied by a release of energy. Research into fusion reactors is advanced, but no commercial reactor has yet been built and fusion power is unlikely to supply more than experimental amounts of electricity before the end of this century. When it does, the potential output will be vast, and the 'fuel' will be obtained from ordinary sea water.

At present all the nuclear reactors in the world exploit the principle of fission.

The mass of any element is determined by the mass of the nucleus of its atoms. An atomic nucleus consists mainly of protons, which have a positive electrical charge, and neutrons, which have no electrical charge. Neutrons have very slightly more mass than protons. An element can exist in different forms, or 'isotopes' if the number of protons in its nucleus remains the same but the number of neutrons changes. This alters the mass of the atom, but not its chemical properties. Hydrogen, for example, the lightest element, has a nucleus comprising one proton; deuterium has one proton plus one neutron, and tritium has one proton plus two neutrons. Most elements exist naturally as mixtures of different isotopes.

Uranium

The nuclei of certain elements are very large, and among those large ones there are some which are unstable: they 'decay'. The most common isotope of uranium, for example, has a nucleus whose mass is 238 times that of a single proton or neutron, and every now and then a uranium nucleus will release an 'alpha' particle, consisting of two protons and two neutrons bound together. Because it loses protons its chemical characteristics change, and it turns into thorium, with a nuclear mass equal to 234 protons or neutrons. Thorium also releases alpha particles, and by the time the decay finishes and the nucleus is stable the resulting element is lead. Uranium, then, decays eventually into lead.

As well as losing alpha particles, a nucleus of uranium-238 occasionally breaks into two smaller nuclei. It is a rare event, happening roughly once for every million alpha decays, but when it happens the nucleus releases some free neutrons, along with other particles, and energy.

Hinkley Point 'B' power station, beside the Bristol Channel, near Bridgwater, Somerset, a 1320 MW Advanced Gas-cooled Reactor using uranium dioxide fuel. Hinkley Point 'A', beside it, is a 500 MW Magnox reactor. The two share fuel handling and control facilities.

Of all the uranium in the world, 99.28% is uranium-238, but mixed with it there is a small amount, 0.71%, of a different isotope, uranium-235. If a neutron moving with not too much energy – called a 'slow' neutron – should collide with a nucleus of uranium-235 it will be captured and held, forming uranium-236, but uranium-236 is very unstable. An impact from one more neutron will cause it to break into two smaller nuclei, of yttrium-95 and iodine-139, with a release of energy. The yttrium and iodine are also unstable and they decay, releasing more energy and more neutrons.

Uranium is distributed very thinly in rocks but if it can be extracted and concentrated the chance increases that the slow neutrons emitted during fission will collide with another nucleus rather than escaping from the uranium altogether. Since each collision releases more neutrons, which also have a chance of colliding with a nucleus, the result is self-perpetuating, a 'chain reaction'. However, the reaction proceeds through the fission of the rare uranium-235 much more than through occasional fissions of uranium-238.

The fission of uranium-235 nuclei releases no more energy than the fission of uranium-238 nuclei. The difference lies in the fact that uranium-238 nuclei release their energy gradually over millions of years; uranium-235 can be induced to release it all at once. For this reason uranium-235 is the most basic fuel for nuclear reactors. The amount of energy is considerable. Uranium-235 releases about three million times more energy than can be obtained from an equivalent quantity of coal.

The proportion of uranium-235 can be increased to about 3% to increase the efficiency of the fuel, and such 'enriched' uranium is used in many older reactor designs.

Using fast neutrons

However, it is possible to extract useful amounts of energy from uranium-238 if its nuclei are bombarded not with slow neutrons but with fast ones, that have been greatly accelerated. The reactor which does this, using fast neutrons, is called a 'fast breeder reactor'. The uranium-238 changes to uranium-239 which undergoes a rapid 'beta-decay' in which a neutron changes into a proton with the emission of an electron and an antineutrino. The mass of the nucleus remains the same but the element changes from uranium to plutonium-239, and when it is bombarded with fast neutrons plutonium-239 undergoes fission readily, producing zirconium-100 and xenon-137, and releasing neutrons and energy. By using both isotopes of uranium instead of only one, the energy obtained from a given amount of uranium is increased by nearly 150 times. Such a reactor is called a 'breeder' because it produces plutonium faster than the rate of plutonium fission, so providing a surplus that can be used in other reactors.

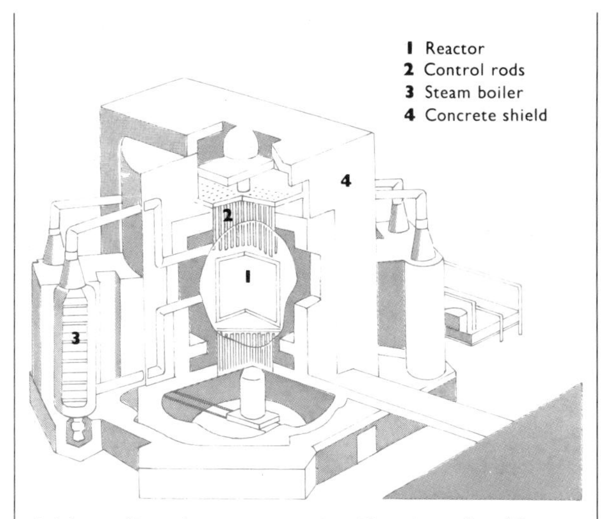

A (thermal) nuclear power station. Heat is produced by radioactive decay in the core of the reactor. The core is cooled, and the heat extracted, by passing water or gas around it. This hot fluid then heats water to produce steam which drives turbines, and the turbines generate electricity.

How a reactor works

Some of the neutrons released by nuclear fission have more energy than others: there are both fast and slow neutrons. If the fuel is uranium-235 the fission reaction is sustained by the slow neutrons. Fast neutrons can be slowed down by making them pass through substances such as graphite, deuterium oxide ('heavy water'), or even ordinary water held under pressure. Such a substance is known as a 'moderator'. The fuel, which varies from one type of reactor to another, is placed in metal tubes and the tubes are lowered into the moderator – into holes if the moderator is solid. The rate at which the reaction occurs is regulated by 'control rods', which can be raised or lowered into the moderator. The control rods are made of a substance that absorbs all neutrons, whether they are fast or slow. The combination of moderator, fuel and control rods is called the 'core' of the reactor.

The fission energy is released as heat, so the reactor is surrounded by pipes through which passes a fluid that carries away the heat, converts into steam water held in a separate system, and the steam drives turbine generators. The fluid that carries away the heat from the core also cools it, and so it is known as the 'coolant'. If the cooling system were to fail the core might overheat. The reaction could be stopped by dropping all the control rods into position, but there is also at least one emergency core cooling system, designed to take over in the event of a failure of the primary cooling system.

Fast breeders

There are something like 250 ordinary fission reactors in the world – nearly 70 of them in the USA and more than 30 in Britain – but very few fast breeder reactors, which are a more recent development. They are more complex technically and use molten sodium as a coolant rather than water or a gas, but they are no more dangerous to operate.

The fast breeder exploits the fission reaction of plutonium-239 made from uranium-238, for which it requires fast neutrons, and so there is no need to slow down the fast neutrons that are emitted. Thus a fast breeder reactor uses no moderator, and apart from an initial charge of plutonium needed to get the reaction started, it can use as fuel the spent fuel from conventional reactors. It is 'spent' because its uranium-235 content has been depleted, so that what remains is about 97% uranium-238 and its oxide, rather less than 1% plutonium (because there is some fast neutron bombardment of uranium-238 even in reactors using a moderator and although much of the plutonium undergoes fission, some remains), and about 2% of other fission products. Because a conventional reactor can use only a tiny proportion of the total mass of uranium it receives, it is very wasteful, and in the long term the economics of nuclear power generation make no sense unless there are fast breeders to use uranium-238 to make plutonium-239 which other reactors can use.

The site of the 250 MW Dounreay prototype fast reactor seen from the Pentland Firth. It began generating electricity in 1975.

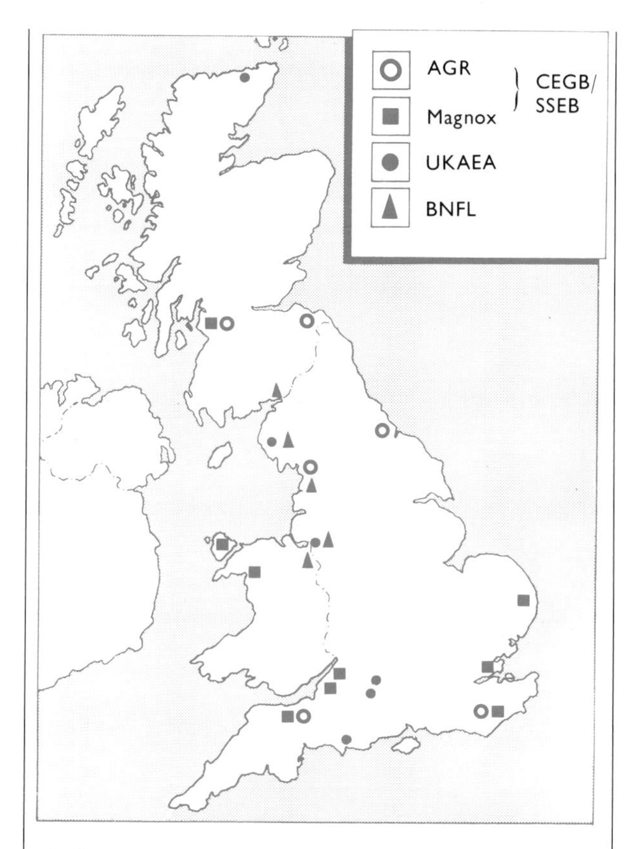

The location of nuclear power stations in Great Britain.

The nuclear fuel cycle

The fuel for nuclear power production is engaged in what is called, rather misleadingly, the 'nuclear fuel cycle'. The name is misleading because the process is linear, not cyclical.

It begins with the extraction of uranium – at present from rocks but in the future possibly from sea water. The ore is crushed and the uranium separated from it in solution. The uranium is then dried and pressed as uranium oxide, known as yellowcake. The yellowcake may be used directly in some reactors, but for others it requires further processing – enrichment – to remove some of the uranium-238 and thus increase the proportion of uranium-235.

Once a conventional reactor has received a full charge of fuel it is maintained by removing some 15 to 20% of its fuel each year and replacing it with new fuel. The spent fuel can be reprocessed for use in a fast breeder reactor, or stored as waste. In Britain the spent fuel is sent to Sellafield, Cumbria, where for the time being it is stored. A plant for extracting uranium oxide from spent reactor fuel and reprocessing it, called the 'thermal oxide reprocessing plant' or THORP, is being built at Sellafield and should be operating by 1990.

Nuclear waste disposal

The final waste, and everything that has come into contact with a reactor core or fuel, is radioactive: it contains isotopes of elements that decay, emitting alpha particles, beta particles (electrons), or gamma (very short wave electromagnetic) radiation as they do so. Waste that is to be stored is classified as low, medium, or high level depending on its level of radioactivity.

Low-level waste is cool and emits very little radiation. It may consist of gases and liquids that remain radioactive for only a short time after which they can be released into the atmosphere, rivers or the sea, and solids that can be incinerated, buried, or sealed in containers and buried or dumped in the deep oceans. Pressure from environmental organizations has led to a ban on dumping at sea, but there is no evidence that such dumping does, or ever could, cause any harm at all. Even if the containers were to leak, the deep ocean floor is a desert and by the time the waste became incorporated in living organisms it would have decayed so much as to have virtually ceased to be radioactive. Wastes that were formerly dumped at sea must now be disposed of on land.

Medium-level waste must be sealed and contained for a long period, until the radiation it emits has fallen to a level that is harmless. High-level waste is very hot, intensely radioactive, and must be stored carefully until its radiation falls sufficiently for it to be classified as medium level.

High-level wastes are stored in liquid form and so far none have been disposed of finally in Britain. At present all of the high and medium level civil and military wastes accumulated since the 1940s are held at Sellafield in about a dozen tanks, and these have never leaked or caused any problems.

In future the liquid waste will be reduced in volume by a vitrification process that converts it into a glass-like substance (a borosilicate glass), although there is an alternative process developed in Australia in which the waste is converted to a rock-like

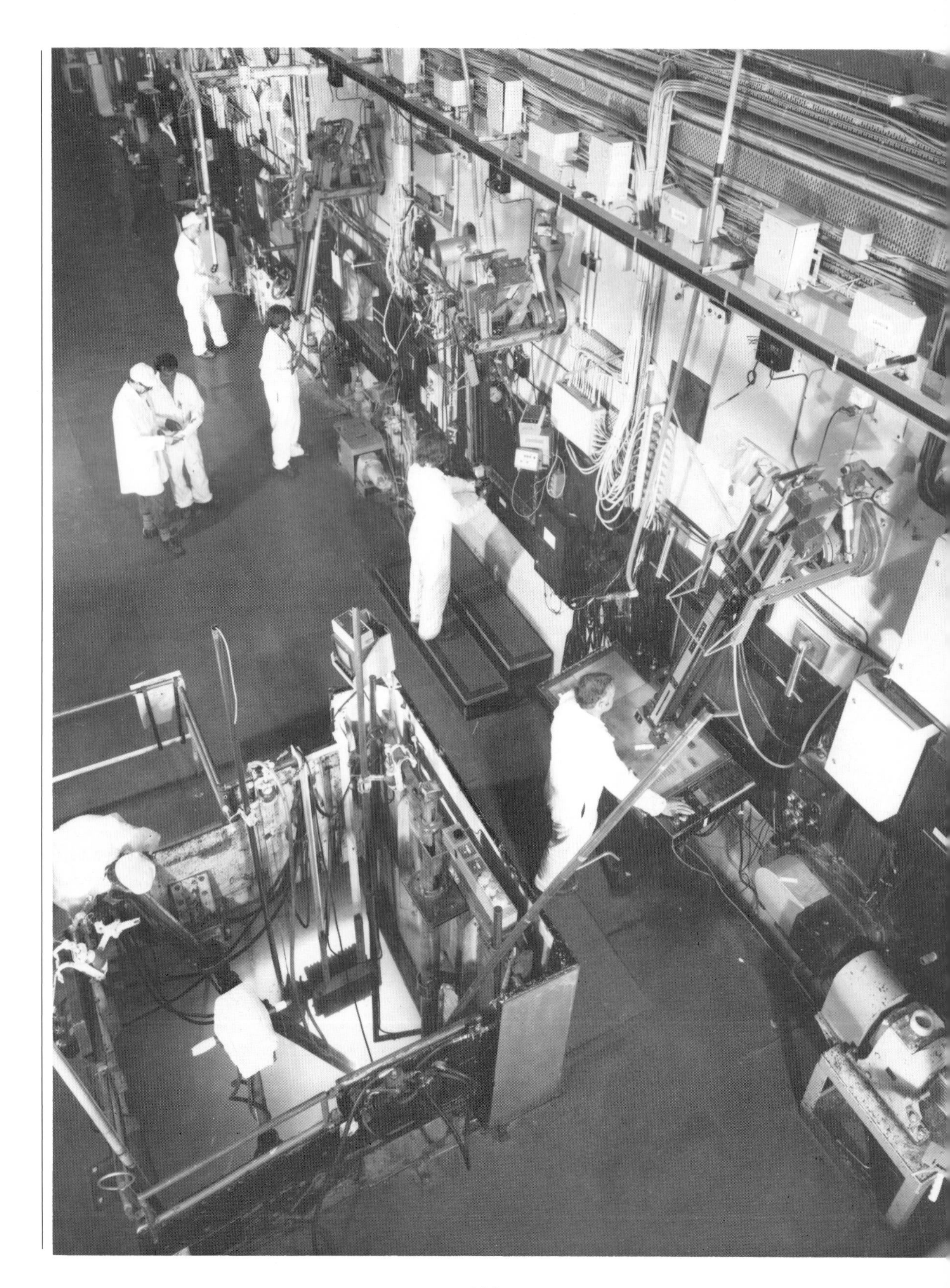

Sellafield reprocessing plant, Cumbria. ***Left*** *Containers being stripped from spent fuel elements under visual inspection inside thick-walled concrete 'caves'. Unused uranium, plutonium and fission products are separated by counter-current solvent extraction.* ***Top*** *transport flask containing spent Advanced Gas-cooled Reactor fuel being lowered into a cell ante-chamber prior to unloading in a storage pond, where it will await reprocessing.* ***Above*** *Flask containing spent fuel, arriving by rail.*

substance, called synrock. Plans have already been made for vitrification of wastes, but the synrock process seems set to take over from the vitrification process eventually. After a very long time it is possible for small quantities of waste to be dissolved out of glass by water. Research suggests that synrock is much more resistant to such 'leaching'.

The quantity of waste is not large once it has been solidified. In the course of its thirty year lifetime a commercial power station reactor might produce enough waste to fill the kitchen broom cupboard and all the wardrobes in a small modern house.

This will be sealed into cylindrical steel tanks, probably about one to 1.5 metres (four to five feet) long and about 15 cm (six inches) in diameter. These will be placed in lagoons of water where they cool quite rapidly.

After about ten years, although still hot, they are cool enough to remove from the water and cover with a thick cast iron casing. In this form they will be moved to an air-cooled cave or a specially built underground store that is well ventilated. The air passing over them cools them, and they will have to stay in this store for 50 to 70 years.

The permanent store

After about 70 years the containers will be cool enough to be handled fairly easily – although they will still be moved only by machines. This is when they will be moved to their permanent storage site. This will be a place underground, or perhaps on or below the sea bed. The containers will be designed to resist corrosion in order to prevent the contamination of water, either by making them from a highly corrosion-resistant material, or by using a material such as iron, which corrodes but at a known rate, so its thickness will determine how long it lasts. If stored on land, the vitreous form of the material will prevent any serious contamination of water for up to 1,000 years should the casing fail, but as an additional precaution sites will be chosen because they are dry and geologically stable, or because like salt and some clays they are soft enough to seal themselves if they are broken into, or because they have a high capacity for absorbing radiation.

The wastes will have to remain in their final resting place for between 500 and 1,000 years. At the end of that time they will be cool, not dangerous to touch or move, but still dangerous if the waste should be broken into small pieces or powder and swallowed or inhaled.

The plans thus allow for secure storage for up to 1,000 years, with the form of the waste itself providing a further 1,000 years safety margin – or perhaps longer if the synrock process is used.

It is true that some of the products of nuclear fission remain highly radioactive for billions of years, but the quantity of these is extremely small, they are mixed thoroughly with other wastes that decay much more quickly, and what matters so far as safety is concerned is the total amount and type of radiation emitted. It is not true, therefore, that wastes must be kept secure for millions of years: 1,000 years with an additional 1,000 year safety margin is ample.

Is it safe?

At the end of that time it will still be radioactive – so will it be safe? It is here that we enter the argument about the dangers of radiation and of the industry as a whole. The aim of storage is to isolate materials securely until they are no more radioactive than the uranium-bearing rocks from whence they came.

In theory, all exposure to 'ionizing' radiation is harmful. This is particle or electromagnetic radiation that can penetrate tissue and impart energy to

Power station radioactive waste can be stored on the surface indefinitely, but if final storage is needed used fuel rods will be solidified into a kind of glass or rock, then placed first in a lagoon of water. After 10 years they will be removed, sealed inside cast iron casings, and moved to a well-ventilated cave or specially built store. They will remain there for 50–70 years after which they are ready for final, permanent disposal below ground or below the sea bed, where depending on the composition of the waste they will need to remain undisturbed for 500–1000 years. By that time they will be no more radioactive than natural uranium-bearing rock.

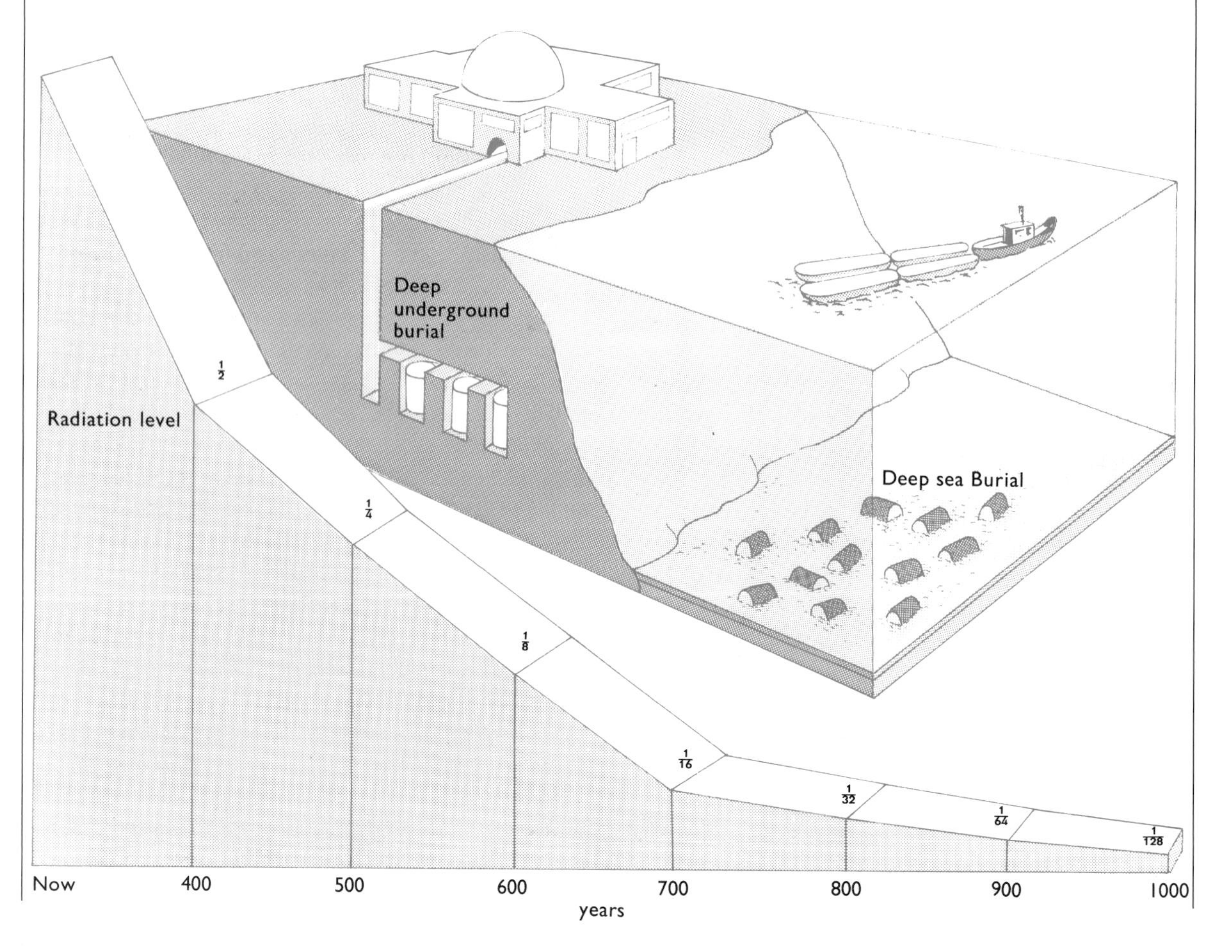

Train crash staged by the Central Electricity Generating Board at Old Dalby, Leicestershire, to demonstrate the safety of spent fuel transport flasks. The flask was not broken.

large organic molecules inside cells, so disrupting them and causing damage or, in the case of nucleic acids, altering their behaviour. Usually such damaged cells die, or are destroyed by the body itself, but it is possible for them to become malignant and therefore cancerous. Having said that, however, it is obvious that there must be a certain amount of radiation we can tolerate without coming to any harm, because we live in a radioactive environment anyway. It seems likely that there is a threshold somewhere above the ordinary level of 'background' radiation below which our bodies cope perfectly well with such radiation damage as they sustain. Above that level there may be injury but the likelihood of it remains low until the radiation dose is quite high.

Radiation is dangerous, but much less so than many people fear. Some forms of leukaemia can be caused by radiation, for example, but leukaemia of all kinds accounts for only about 0.5% of all deaths in the United Kingdom. Road accidents account for 1.2%. Plutonium is extremely poisonous if you should swallow or inhale particles of it, but it is a 'beta-emitter', releasing electrons which cannot penetrate tissue very far, and so it is not especially dangerous to handle.

The most serious accident ever to have involved a nuclear reactor occurred at Chernobyl, Russia, in April 1986, when one of a complex of four reactors of a Soviet design not used in other countries was destroyed in a series of explosions. Two workers were killed immediately and a further 29 died later, from injuries or radiation sickness, and fall-out contaminated a wide area of Europe.

Although restrictions were imposed on the consumption of contaminated produce, in Britain the threshold level beyond which these measures were applied was set between one-hundredth and one-tenth of a safe dose agreed internationally only a few weeks before the incident. The health of people who really were affected by fallout, in the USSR, was monitored closely. Two years after the event there was no evidence that their health had suffered, although it is quite possible that the lives of as many as one thousand Soviet citizens may have been shortened by a small amount.

The notorious Three Mile Island accident in America released so little radioactive material that no one was harmed by it.

Perhaps we should consider the safety of the alternatives. Some coal-fired power plants, for example, emit up to twelve times more radiation than a nuclear plant of the same size. The burning of oil and gas releases carbon dioxide, and when coal burns a range of substances is released, many of which are harmful and some of which are known to cause cancer. Before its use was banned, coal burned in ordinary domestic fireplaces rather than power stations is known to have killed thousands of people in London. It can hardly be called safe.

There could be an accident, but the chance of one that leads to a substantial release of radioactive materials is extremely small.

'The China syndrome'

In the worst possible reactor accident the core cooling system fails, the emergency system also fails, the reactor fails to drop all its control rods into place automatically and so shut down the reactor as it is meant to do, and the core overheats until it melts. This is the so-called 'China syndrome', in which the whole reactor melts, the floor beneath it melts, and the intensely hot, intensely radioactive mass of molten material descends (supposedly in the general direction of China via the centre of the Earth, and hence the name) until it meets the water table, where it reacts violently with the water and explodes, blowing up the entire power plant and voiding radioactive wastes over the surrounding countryside. It has been examined as thoroughly as it can be examined without actually making it happen (although plans were made to carry out that experiment, but they were abandoned because of the cost) and in reality the 'melt-down' would be held by the floor of the reactor building, and the material would cool rapidly rather than continuing to heat. The China syndrome is a myth.

It has been said that a direct hit on a nuclear power plant by a nuclear weapon would cause a major release of radioactive material – but an attack with a nuclear weapon implies nuclear war, and radiation hazards due to wrecked power plants would be the least of our problems.

Nuclear power generation is not perfectly safe – nothing is – but although some members of the public may have been injured by it there cannot be very many, and there is no conclusive evidence that any have been. It causes less damage to living organisms than its principal alternatives, requires less space for the transport and storage of fuels, and generally has a much smaller adverse effect on the natural environment than is caused by the burning of fossil fuels.

Solar energy

Sunlight and solar heat can be used to generate electricity or for direct heating, but they are limited geographically. In Britain, for example, there may be too little warmth in the sunshine to provide useful amounts of heat in winter, although it can contribute to water heating in summer. Unfortunately the installation itself is expensive, and because heat is transferred by warming water the plumbing is complex and needs maintenance.

Sunlight can be used directly by means of cells containing substances that emit electrons (electricity) when they are bombarded with photons (light). The output is very small but could be useful provided a large enough roof area is available to catch the light. The cost of electricity generated in this way is higher than that generated conventionally on a large scale.

Wind power

Wind power is used already on a limited scale, but probably could not be used on a large scale. The largest wind generators produce about one megawatt (one million watts). Beyond that size they start to suffer from grave engineering problems. If they are kept small enough to work reliably probably a battery of about one thousand would be needed to match the output from a conventional (one gigawatt) power station. Although their output is small, wind generators themselves are large, and must be sited in the most exposed, and therefore most highly visible places. Wind power generation is useful in isolated locations, but it is no cheaper, and probably more expensive, than conventional generation.

Waves and tides

Wave-power generators have also been designed but they, too, are large and can be used only off those coasts where the sea swell is sufficient. In Europe this confines them to the Atlantic coasts. Where the tidal flow is large it can be harnessed. The inflowing water is held behind a barrier, possibly turning turbines as it enters. When the tide ebbs so the water behind the barrier is at a higher level than the water in front of it, the barrier is opened and the water turns turbines as it leaves. Such tidal barrage schemes are possible only in a few places. There is one at La Rance, in Brittany, and one has been planned for Britain, in the Severn Estuary near Bristol. Tidal barrages are expensive to build but then operate cheaply and can produce large amounts of power. Environmentally they are less neutral than they may seem because they alter the system of currents and therefore the deposition of mud on which the estuarine ecosystem depends.

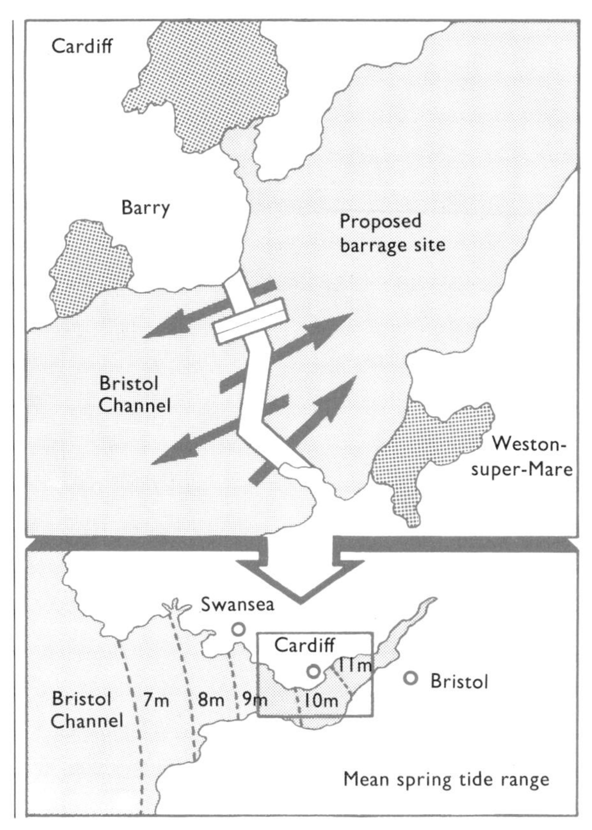

Right *Proposed tidal barrage scheme for the Bristol Channel.*

The Severn bore. Twice a day the incoming tide sends a wave 1.5m high as far as Newnham, Gloucestershire. With one of the world's largest tidal ranges the Severn offers one of the few practicable sites for a tidal barrage.

Opposite page *The Tucurui Dam on the Rio Tocantins, Brazil, is one of 30 planned in the Amazon basin. They will flood a large area of forest.* **Left** *Altamont Pass wind farm, California, where 300 turbines generate about one-quarter of the electricity produced by a modern power station.* **Below** *Solar panels being used to augment conventional water heating in Dulwich, London. In warm weather they can save fuel, but they are not cheap to install.*

Large dams

Hydroelectric power, from the flow of river water, is widely used. It can be exploited on a large scale only where the rivers are suitable and usually requires large dams. The dam forms an artificial lake, and it is the flow from the lake that turns the turbines to generate the power. A large dam is one which is more than 150 metres high, or with a reservoir whose volume is more than about 25 billion cubic metres. The first large dam was the Hoover, on the Colorado River, built in 1936, and at present there are 175 large dams in the world, with a further 38 under construction and more planned. California has almost one-tenth of the total and there are others elsewhere in the United States, Canada and the USSR, but many are in Third World countries and that is where many new dams are being planned. The Tucurui, for example, in the Amazon Basin of Brazil, is already one of the largest, and it is the first in a programme of more than 30 dams to be built along rivers in the Basin.

Dams are not new, but modern large dams are not like traditional dams and to some extent they are still experimental. If you multiply the number of nuclear power plants in the non-communist world by the total number of years they have been in operation the answer is in the region of 2,000 'reactor-years'. If you multiply the number of large dams by the number of years since they were built the answer is about 1,200 'dam-years'.

Already we know that large dams are not without their dangers. While the lake behind the Hoover dam was filling there was an earthquake, and between the 1930s and 1960s there were earthquakes in the neighbourhood of four more large dams. Water is heavy and the lake behind a large dam weighs billions of tonnes. If the rocks below are faulted the weight of the water may be enough to cause two large masses of rock to move in relation to one another: that is an earthquake.

The construction of any artificial lake involves flooding land, and on a large scale the ecological consequences can be grave. In the tropics and subtropics there may also be adverse effects on human health. Schistosomiasis (bilharziasis) is a debilitating disease caused by a blood fluke, any one of several species of parasitic trematode flatworms. All flukes have a life cycle involving more than one host and in some species humans are the only final hosts; other species can also parasitize monkeys or other mammals. The intermediate host is an aquatic snail, and during a pronounced dry season many of the snails die. Dams hold water all the time, so providing a suitable habitat for snails and allowing the flukes to increase, so that dams, and especially large dams, can and do lead to an increase in schistosomiasis which then requires – but often fails to receive – treatment.

Oasis on an oxbow bend in the River Indus, Pakistan. Large rivers deposit silt as they advance across their floodplains. Such soil, the result of natural processes, is rich in nutrients.

A dam allows water to pass under control, but by holding it before it is released much of the silt it carries settles to the bottom of the lake. Eventually the silt fills the lake and because the lake itself is usually very deep, dredging is often impossible and the dam reaches the end of its useful life. Meanwhile the part of the river downstream of the dam receives less silt than it did. River silt is rich in plant nutrients. If the region downstream is farmed in most cases the farmers will have relied on the silt, which will have been delivered free of charge during periodic – or in some cases annual – floods. Without the silt they will have to buy fertilizers, and so the cost of their farming increases. If the region is not farmed, its ecology will change as the amount of nutrient and water reaching it change.

Combined heat and power

There are no easy answers to satisfying our need for energy, but there are ways to make economies. Already, since the first rise in oil prices during the 1970s, we have learned to insulate our buildings more efficiently and to reduce waste. We could do more, especially if we were to make better use of our large power stations.

Conventional power stations produce heat which is used to produce steam and the steam drives turbines. The overall efficiency of the operation is around 35 per cent because the fluid that carries the heat away from the furnace or core is cooled, usually in large towers, so that of all the heat produced only about one-third ever reaches the consumer as useful energy. The rest is lost into the air or nearby water. If, instead, this hot fluid were cooled more slowly by passing it through buildings nearby, those buildings and their water supply could be heated at very little cost and the efficiency of the operation could be increased to about 70 per cent or even more. Such 'combined heat and power', or 'CHP' schemes have been installed in a few places. They require that power stations be built in urban areas, and they may have to be smaller power stations than most of those in operation at present because very large power stations produce more heat than can be used profitably in a CHP scheme, but the idea has many supporters.

A combined heat and power scheme uses cooling water from a power station to provide water and space heating to nearby buildings, so increasing dramatically the efficiency with which the power station uses fuel.

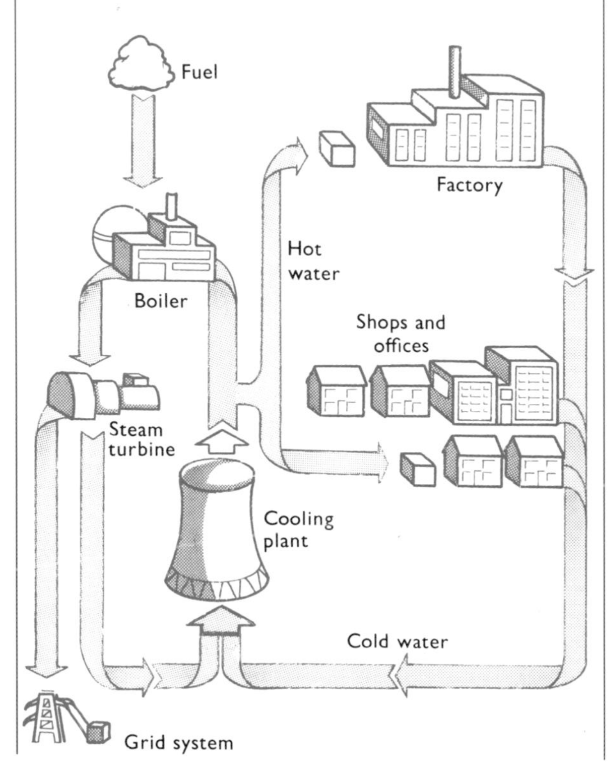

What of the future?

In years to come it seems likely that oil will be abandoned as an important fuel for power generation and that we will rely less than we do now on gas and coal – although coal will remain in use for a very long time. Wind generators may provide some power in remote places, but in high latitudes it is doubtful whether solar heating or electrical generation, or systems based on the movement of waves will contribute much to total power output. Tidal barriers may be built in the few places suitable for them, and the Severn Barrage – the only suitable British site – could produce as much as four or five conventional power stations. Dams will continue to be built, but again the number of possible sites is limited. It is possible that as new power stations are built to replace old ones at least some will incorporate CHP schemes.

Nuclear power, with fast breeder reactors and eventually thermonuclear fusion reactors, will be increasingly important. Its expansion will reduce environmental pollution and damage to landscapes.

If you wish to encourage the development of more sensible combined energy and environmental policies you may need to start your own organization. It should encourage energy conservation, but with caution because in buildings conservation usually involves reducing the frequency with which inside air is exchanged for fresh outside air. Building materials release radon, a radioactive gas, which is removed quickly enough when air changes are frequent, but reducing the rate of air exchange can allow radon to accumulate. For environmental reasons you should favour nuclear power. The existing environmental organizations will not support such policies, but if you form a new organization it might attract much popular support.

CHAPTER 9

Whales, seals, sea otters and 'mermaids'

It does not follow that people who are concerned about the natural environment are also interested in conservation, although they are almost bound to be sympathetic to its aims. The reverse is not necessarily true. Conservationists do not have to be environmentalists and often they are not, at least in the eyes of environmentalists themselves. It is easy to confuse the two because members of one movement are always ready to use arguments borrowed from the other to support their case on particular issues. The two movements overlap at many points but they remain separate because their central philosophies are distinct. The conservationist is concerned about the fate of non-humans and seeks to protect them and their habitats. The environmentalist is concerned about humans and the human habitat, which may not be the same thing.

From time to time, however, the two movements merge, and the most spectacular and profitable of such mergers led to the campaign to end commercial whaling. Such whaling will end in 1988 in every country except for the USSR and Norway, and they are likely to follow. Ostensibly whaling will cease for a five-year period during which its effect on whale populations will be assessed more accurately than has been possible in the past. In practice, a five-year halt to whaling almost certainly means that whaling will not resume. The ships will have found other uses, or will have been scrapped, the workers will have found other employment, and substitutes will have been found for the whale products that are the commercial basis of the industry. The campaign

The whole of a humpback whale (Megaptera novaeangliae), *photographed as it breaches. Humpbacks also roll over an over in the water, and swim on their sides. They grow up to 15m long. Their mouths contain 300 to 400 baleen plates.*

succeeded, but it took a long time, a great deal of effort, and it was expensive.

The collaboration began when environmentalists took up a conservationist cause. They were able to contribute – and to develop for their own use in other campaigns – political toughness and sophistication, together with skill at attracting the attention of the press and television and at presenting their case simply and emotionally, and so winning widespread popular support. The conservationists contributed their scientific knowledge and, no less important, the names of their own supporters, a list that included internationally respected scientists, and in some cases famous ones, and leading public figures. An advertisement in *The Times* (London) on 25 June 1973, headed 'One is killed every 20 minutes. Is this carnage really necessary?' demonstrated the emotional appeal, and its signatories gave it suitable weight. They included HRH Prince Bernhard of the Netherlands, HRH The Duke of Edinburgh, Commander Jacques Cousteau, Sir Frank Fraser Darling FRSE, Dr Thor Heyerdahl, Sir Julian Huxley FRS, Prof Claude Levi-Strauss, Dr Konrad Lorenz, and Sir Peter Scott.

The collaboration was successful, but its aims remained distinct. The conservationists sought to protect the whales from a level of exploitation that for all the scientists knew might so deplete their numbers as to bring some of them to extinction. They wanted to save whales because whales should be saved. The environmentalists had wider aims. They wanted to alter human behaviour for the sake of humans. The Friends of the Earth *Whale Campaign Manual 2*, published in 1974, stated:

'Man, as the most successful of all species, has special responsibilities. The decisions he takes now affect not only his own and his children's future, they also affect that of many other species. In this

The mouths of whales. Baleen whales have very large mouths, containing large plates suspended from the roof. The whale takes in a mouthful of water, lowers the plates, then uses its tongue to force the water out again past the plates, which catch food items. The toothed whales have simple teeth, the number varying widely according to the species.

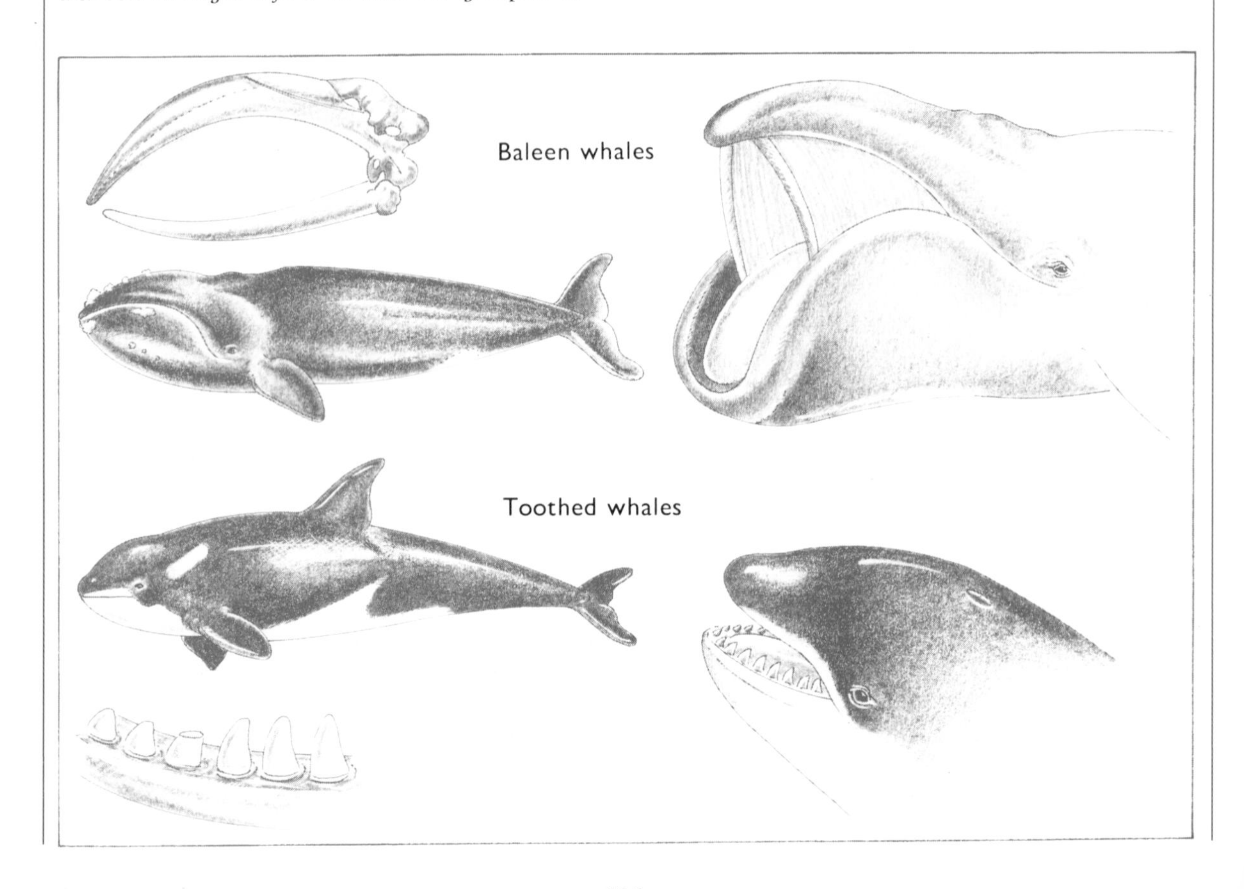

situation the over-exploitation of whales becomes as immoral as it is insensitive.'

They wanted us to be better people. The whales are probably secure. Whether we have become better people is debatable.

Who are the victims?

Vertebrate animals first evolved in the sea and it was not until about 350 million years ago that some of them began to live partly on land. The first mammals did not appear until about 150 million years after that, the land-dwelling descendants of land-dwelling ancestors, but by around 40 million years ago one group had returned to the sea, and fed on fish. The first of the archaeocetes, ancestors of the modern whales and porpoises, were about the size of the smaller modern porpoises, had long snouts, and their teeth resembled those of early carnivores.

Today there are about 84 species of whales, grouped together in the order Cetacea which is divided into two suborders, the Odontoceti comprising whales which have teeth as adults, and the Mysticeti, whose members lack teeth as adults but have baleen plates. Most cetacean species are toothed, and belong to the Odontoceti, but the largest whales, and the largest animals of any kind ever to have lived on Earth, are mysticetes. The porpoises are odontocetes. So are the only freshwater whales – four species of fish-eating river dolphins, three of which are found in the rivers of India, the fourth in the Amazon and Orinoco rivers of South America, and the eight species of the family Stenidae. These are porpoise-like animals, about 2 m (7 ft) long, found in warm-temperate coastal waters and in the rivers of tropical South America, Africa and Asia. *Sousa*, known variously as the white, hump-backed, or long-backed dolphin, feeds on plants and may be the only vegetarian whale.

Cetaceans are mammals, but they have adapted to life in the water and never move on to land. They breathe air with lungs rather than gills, but their nostrils have moved to the top of the skull, the 'blowhole', and can be closed. The lungs are very elastic. A whale can exhale a large volume of spent air and inhale a new supply very quickly, and when it dives the heart-beat slows and there are probably other ways in which oxygen is conserved. The hind limbs have disappeared altogether, except for a few

Southern right whale (Eubalaena australis) *displaying the baleen plates attached at the roof of the mouth, through which the whale strains water to collect food.*

small bones that are what remains of the pelvis, and the forelimbs are modified into paddles that can be moved only from the shoulder. The digits are still present, but small and hidden. Some whales have a dorsal fin, made from a fold of skin. The tail has become 'flukes', resembling the tail-fin of a fish but mounted horizontally rather than vertically as it is in all fish. The body is rounded, but is more rigid and bulky than that of a fish and a whale propels itself with only its tail – it cannot make fish-like, sinuous movements of its whole body. Young are born in the water, and must be raised to the surface to breathe. Whales have lost most of their body hair, but keep warm by means of a layer of fat – blubber – just beneath the skin. The blubber layer is about 2.5 cm (1 in) thick in porpoises, but can be 30 cm (12 in) thick in large whales.

The odontocetes have simple, peg-like teeth, usually in both jaws, and probably used only to hold prey. The 6 m (20 ft) long male narwhal (*Monodon*) of Arctic waters has a straight tusk with a spiral groove up to 2.7 m (9 ft) long, actually an overgrown tooth, whose purpose is unknown.

The mysticetes have huge mouths and strong, muscular tongues. Vestigial teeth are present in the embryo, but in adults comb-like strips of keratin – dead skin – called baleen or whalebone, more than 3 m (10 ft) long in large whales, hang from the roof

Above *Dolphins are intelligent, gregarious and playful. These are swimming in the eastern Pacific.* ***Left*** *The killer whale* (Orcinus orca) *is a fierce predator, but probably much less dangerous to humans than used to be supposed.* ***Right*** *Walruses live in Arctic waters, and except when mating in separate male and female herds. They use their tusks to haul themselves out of the water, as weapons, to rake the sea bed for food, and to keep breathing holes open in the ice.*

of the mouth and have jagged, fringed rear edges. The whale opens its mouth, takes in a mouthful of water, pushes the water out past the baleen plates, then licks the food it has trapped from the back of the plates. The baleen whales are very large indeed, but feed exclusively on very small prey.

All the whales are highly vocal. Many use echolocation to find prey, and some, including porpoises, killer whales, and possibly sperm whales, use intense sound to stun or even kill prey. Sound is also used for communication, in the case of some of the large whales often over great distances.

Whales appear to be intelligent and some species are certainly playful. Sick or injured individuals have been seen to receive assistance from other members of their group and many of the smaller whales react to the proximity of humans in a way the humans describe as friendly. These attributes have been emphasized strongly in the campaigns to end whaling.

The cetaceans are not the only mammals to have moved from land to water. At about the time the archaeocetes were establishing themselves at sea, a group of swamp-dwelling subungulates – the group which also includes the elephants – was doing the same, and its descendants became the sirens, mermaids, or sea-cows, comprising the order Sirenia. They earned the name 'mermaid' partly because the females suckle their young at pectoral teats, as humans do. They live in estuaries and

Whales. Comparative sizes

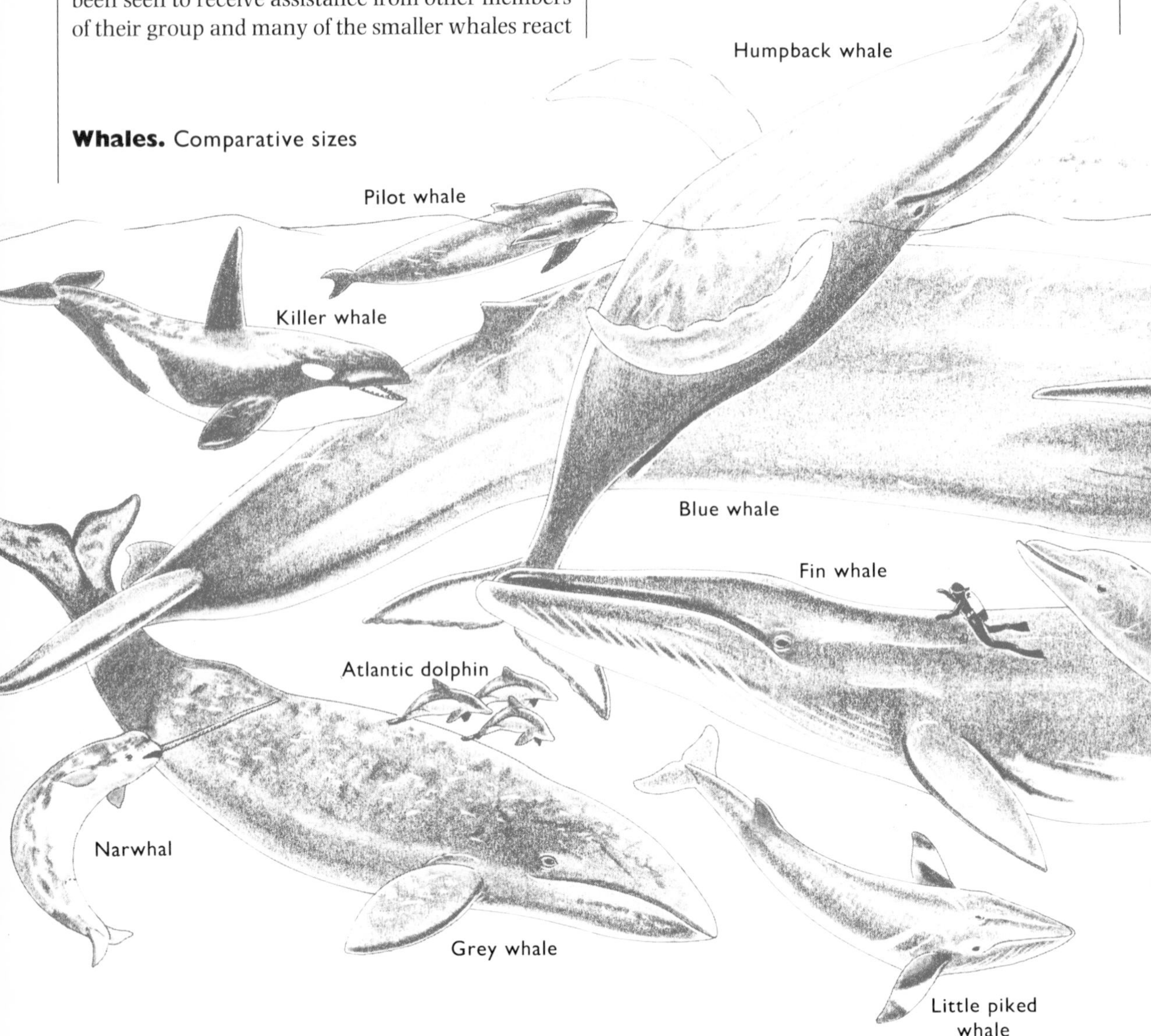

shallow coastal waters, so that a vessel trying to approach them might well run aground, which may account for legends of them luring sailors to disaster and children to watery graves.

The modern sirenians resemble whales in certain respects, but are smaller and exclusively herbivorous. There are two groups of them which differ in several ways. Three species of manatees live along the coasts of tropical Africa and America and one marine species, the dugong, lives in the Indian and Pacific Oceans from the Red Sea to Australia and Taiwan. Steller's sea cow once lived in the Arctic, but the last one was seen in 1768.

Seals are less fully adapted to an aquatic life than cetaceans or sirenians, but their adaptation is graded. There are three groups: the true seals, sea lions, and walruses. The true seals are the most aquatic. The limbs are flippers and the hind limbs point to the rear and cannot be moved forward, so seals are very clumsy on land. They have no external ears. Sea lions do have small external ears and they can move their hind limbs forward, so they get about much better on land. The walruses can bring their hind limbs forward, but have no external ears.

The sea otter (*Enhydra lutris*), which lives among beds of kelp off the Pacific coasts of North America and the USSR, rarely leaves the water, but is less adapted to an aquatic life than other marine mammals. Its limbs are still those of a land mammal, although the digits are webbed, and it has thick fur rather than blubber.

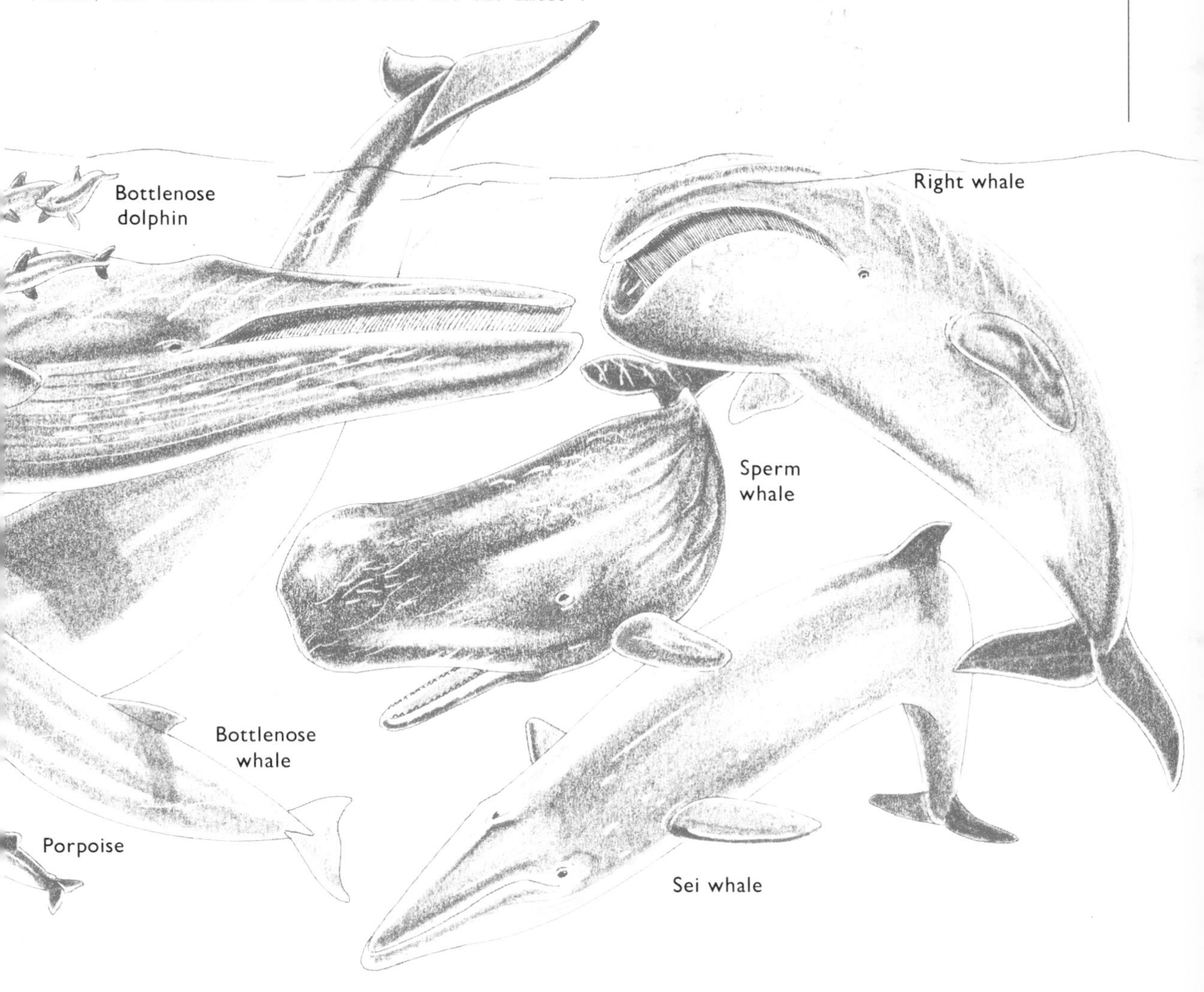

Left The sea otter (Enhydra lutris) *anchors itself in the kelp so it will not drift while it sleeps. It uses rocks to open clams.* ***Above*** *The manatee* (Trichechus mantus), *swimming here in Crystal River, Florida, is a gentle vegetarian.*

The hunt

Killer whales hunt other whales, but otherwise marine mammals have few enemies apart from humans. Unfortunately, humans are formidable opponents.

Dugongs and manatees have been hunted for centuries for their meat, which is said to be excellent. Steller's sea cow was hunted to extinction, but local hunting had little effect on the dugong and manatee populations. Today they are vulnerable to the destruction of their habitat, and to incidental events of which they are innocent victims. The dugongs of the Persian Gulf, for example, may be suffering as a result of the Gulf War. American manatees are protected, however, and the value of sirenians for clearing blocked waterways of plants as well as a source of food has led to suggestions that they should be domesticated fully and farmed.

Sea otters were hunted almost to extinction, for their fur. In 1856 the Russian-American Company marketed 118,000 pelts but by 1885 the catch had fallen to 8,000 and by 1910 it was down to 400. Then hunting ceased because the animal was presumed to be extinct. It was not, and gradually its numbers have increased, helped by the fact that it is protected by the laws of the countries in whose waters it lives.

Most seals are still hunted, or have been, for their skins or fur, and their populations have been reduced drastically. In the late 1960s, for example, there were probably 1,000 to 1,500 Mediterranean monk seals (*Monachus monachus*). In 1985 there may be no more than 30. In some cases traditional hunting is allowed to continue as a means of controlling the size of local seal populations – 'for their own good' – the pressure to continue culling coming from the hunters themselves and from fishermen who believe the seals consume valuable fish, although it is unlikely that seals affect fish populations significantly. Like freshwater otters, it seems more probable that seals catch mainly those fish that are less agile, owing to old age or illness.

Porpoises and dolphins are seldom hunted deliberately, but in some parts of the world fishermen who come across them will kill them because, like seals, they are believed to compete for fish. They are largely protected from such deliberate killing by European fishermen for purely superstitious reasons. European fishermen regard porpoises as in some sense 'human-like', and killing one, even accidentally, is believed to bring misfortune. They are likely to be killed inadvertently, all the same. Drift nets made from very fine fibres and intended to catch squid may ensnare them, and they may be trapped inside purse seine nets when the shoal of fish among which they were hunting is caught. If they are trapped in this way they will be unable to reach the surface to breathe and so they will drown.

The real hunt, the serious one, has been directed against the animals with the greatest commercial value – the large whales.

Whaling

Humans have always hunted whales that came within their reach, but by the ninth century the Basque people had made whaling into a profession. Schools of *Balaena glacialis*, an 18 m (60 ft) long animal when adult, used to swim close to their shores, along the Bay of Biscay. They swam slowly, so were not too difficult to chase in small boats and to harpoon and, unlike most baleen whales, they floated when they were dead, so they could be

The culling of seal pups in Newfoundland has earned worldwide condemnation. Separated from their mothers, the cubs are helpless, and are clubbbed to death. Clubbing kills quickly and accurately, without damaging the fur, but why kill the seals at all?

hauled ashore. For these reasons the species was known as the 'right' whale – it was the right one to hunt. There are several 'right' whales. This one was the Biscayan, or North Atlantic right whale. When it became a rare sight the Basques built better boats and went further in search of it, eventually hunting – and also fishing – as far away as Iceland and Newfoundland. The British and Dutch picked up the habit in the seventeenth century, and so the whaling industry was born.

In America, Indians had also been hunting whales, the humpback (*Megaptera novaeangliae*) and Californian grey (*Eschrichtius glaucus*) as well as the Biscayan right, and the story goes that as, like the Basques, they found themselves voyaging into ever more distant waters in search of their quarry, one day someone killed a sperm whale (*Physeter catodon*) by mistake. That started the US sperm whale industry which followed the pattern of all whaling. The whalers found themselves having to travel further. Then the Californian gold rush, the Civil War, economic depressions, the partial disappearance of sperm whales, and finally the discovery of oil, which was cheaper than sperm oil and for most uses as good, dealt mortal blows to the industry. America has had no whaling ships since 1925, when the last two abandoned the business.

In the eighteenth century, convict ships bound for Australia from England carried whaling gear and went whaling after they had discharged their human cargoes.

The Norwegians went to Antarctica to hunt seals, found whales as well, and were the first to establish a shore station for processing whales, at Grytviken, South Georgia, in 1904.

The Japanese hunted whales on a small scale in Neolithic times but whaling proper did not begin until our seventeenth century, first with harpoons but later with nets.

In addition to 'right' whales there were also wrong ones - whales that swam too fast to be caught or whose bodies sank when they were killed, so they

Above *A modern harpoon gun. Most whales are killed by a harpoon carrying an explosive charge, which strikes the animal with enough force to pierce the skin, and detonates inside the body, causing massive, but not necessarily immediately fatal, internal damage. Death is slow and agonising.* ***Left*** *Two whales that have been killed, strapped to the side of the whaling ship by their tails. They will be towed to a factory ship where the carcasses will be processed at sea.*

could not be towed. New, faster ships made it possible to hunt them, but the 'wrong' whales still could not be killed, not by harpoons thrown by hand. The harpoon gun was invented by a Norwegian, Svend Foyn, in 1803, but it was not until 1868 that it was used successfully for the first time. It could kill even the largest, strongest whale, and Foyn solved the remaining problem of the sinking carcase by pumping it full of air, which made it float.

By the early years of this century steam-powered steel ships with harpoon guns were hunting whales of all species all over the world and in 1905 the first factory ship was working around the Falklands-Malvinas, accepting the carcases from a fleet of catchers and processing them at sea. In the 1930–31 whaling season 42,874 animals were killed, and from 1934 until the big decline the annual world catch exceeded 30,000 animals.

The products

Whales were valuable. It was worth voyaging halfway round the world to find them, and a list of some of the commercial uses to which parts of them were put shows why.

Ambergris (from sperm whales) was used as a fixative for perfumes in cosmetics. Baleen (from any of the baleen whales) was used as 'whalebone' in corsets, whips, umbrellas, and brushes. Whale blood, an incidental by-product, was used in fertilizers and as an ingredient of the adhesives used to make plywood. Gelatine was extracted from bones and cartilage. Hormones were extracted from glands. Vitamin A was obtained from the liver. Skin was used like leather, for handbags, shoes, and bicycle saddles. Sperm oil (or spermaceti, from sperm

A whale being hauled aboard a Peruvian factory ship, to the waiting flensers. They will remove skin and blubber, using razor-sharp knives shaped like hockey sticks. The blubber will be boiled in the vats in the background.

whales) was used in a wide range of cosmetics and toiletries, for dressing animal hides, in many high quality lubricants, plastics, detergents and other goods. Other whale oils were used in soaps, margarine, cooking fats, varnishes, printing ink, dynamite, linoleum, and in many other industrial goods. Tendons were used to string tennis rackets and for surgical stitching. Bone was used as bonemeal fertilizer. Finally, the meat was eaten by humans, but was also used in canned pet foods, in commercial livestock feedstuffs, and sold to mink farms, and zoos. Teeth were made into ivory carvings, known in America as 'scrimshaw', and piano keys.

The long list of useful products – and this list is by no means complete – meant whaling was profitable, but it did not mean it was vital to the economies of the nations engaged in it.

When the whale catch began to fall and whale products became more expensive, substitutes were developed fairly rapidly.

Conservation

If all is well and a species lives in balance with its environment its population may fluctuate widely in the short term but should remain stable over a long period. If conditions deteriorate for the species, deaths may exceed births (there is natural mortality); if conditions improve births may exceed deaths (there is recruitment to the population). In either case other mechanisms may come into play to minimize the change. If there is natural mortality, more young may survive; if there is recruitment there may be less food for older and sicker individuals, so more of them may die. Eventually numbers will stabilize again. Hunting by humans causes natural mortality, and may produce compensatory effects, so calculating the effect of hunting on a species is a complicated business.

The first known attempt to regulate whaling is British and dates from 1324 and is still in force. *De Praerogatorium Regis* (Concerning the Royal Prerogative) declared that all whales, dolphins, porpoises and sturgeons were henceforth 'royal fish'. Any that were caught or found washed ashore had to be presented to the sovereign or in certain cases to the lord of the manor – and they still are, although the bottle-nosed whale (*Hyperoodon ampullatus*) and pilot or caa'ing whale (*Globicephala melaena*) are not 'royal fish' in Scotland. Originally the law regulated the hunting of these species, since only the sovereign could benefit. Today, in the case of whales, the royal prerogative means the Crown (in fact the Department of Trade) is responsible for burying corpses. Sturgeons may be accepted at the palace kitchens if they are fresh.

Further conservation attempts were made during the nineteenth and early twentieth centuries by Russia, Norway and Britain. In 1935 the League of Nations set limits to the size of catches, and in 1937 an International Whaling Convention was signed, imposing a number of regulations intended mainly to prevent the killing of calves and their mothers. Not all whaling nations obeyed the regulations, and some, including Japan, did not even sign the Convention.

The Second World War halted whaling, but a world shortage of oils encouraged it to start again in 1945. In 1944 Norway, a few of whose ships were working out of British ports while their own country was under German occupation, and the English-speaking nations had agreed that the 1937 Convention should be re-activated, but that catch quotas be set based on the amount of oil whales of each species were presumed to yield. This led to the 'blue whale unit', according to which one blue whale yields as much oil as two fin whales, 2.5 humpbacks, or six sei whales. Whaling quotas were calculated using the blue whale unit until 1972, when the unit was abandoned.

In December 1946, under UN auspices, most of the leading whaling nations – but not at first Japan or the USSR – formed the International Whaling Commission (IWC). It was empowered to determine and set quotas, but there were two major difficulties, quite apart from those involved in policing the system.

No nation was obliged to join the IWC and non-membership did not debar whaling. Thus a nation that disagreed strongly enough with IWC regulations could leave, and to this day some whaling nations do not belong to the IWC, although their catch is very small.

The second was the '90 day rule'. When quotas were agreed and set at a plenary session of the IWC, amid full publicity, they were not necessarily final. Any member could lodge an objection within 90 days, and by doing so it would cease to be bound to the agreed quota.

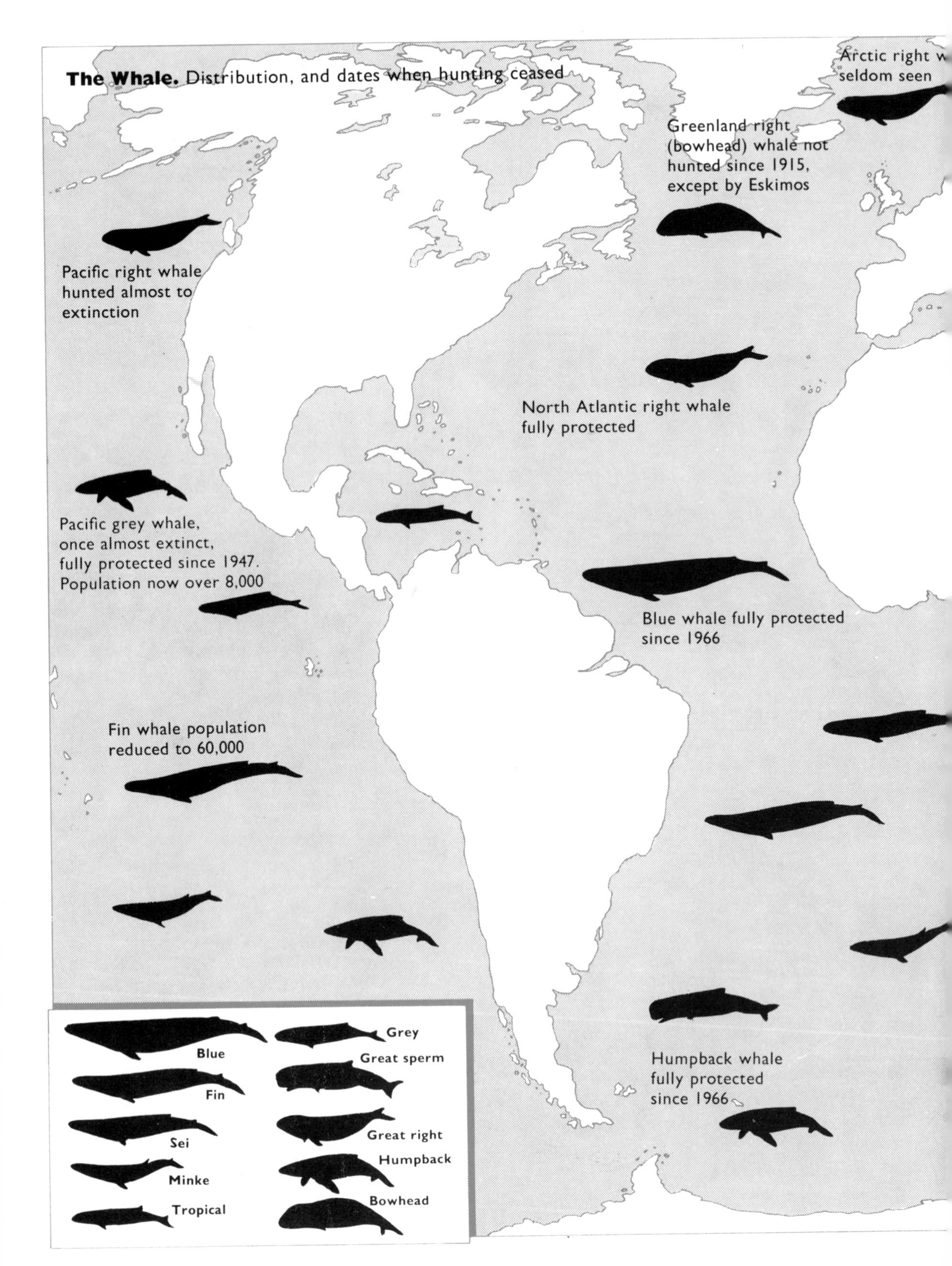
The Whale. Distribution, and dates when hunting ceased
Arctic right w
seldom seen
Greenland right (bowhead) whale not hunted since 1915, except by Eskimos
Pacific right whale hunted almost to extinction
North Atlantic right whale fully protected
Pacific grey whale, once almost extinct, fully protected since 1947. Population now over 8,000
Blue whale fully protected since 1966
Fin whale population reduced to 60,000
Humpback whale fully protected since 1966
Blue
Fin
Sei
Minke
Tropical
Grey
Great sperm
Great right
Humpback
Bowhead

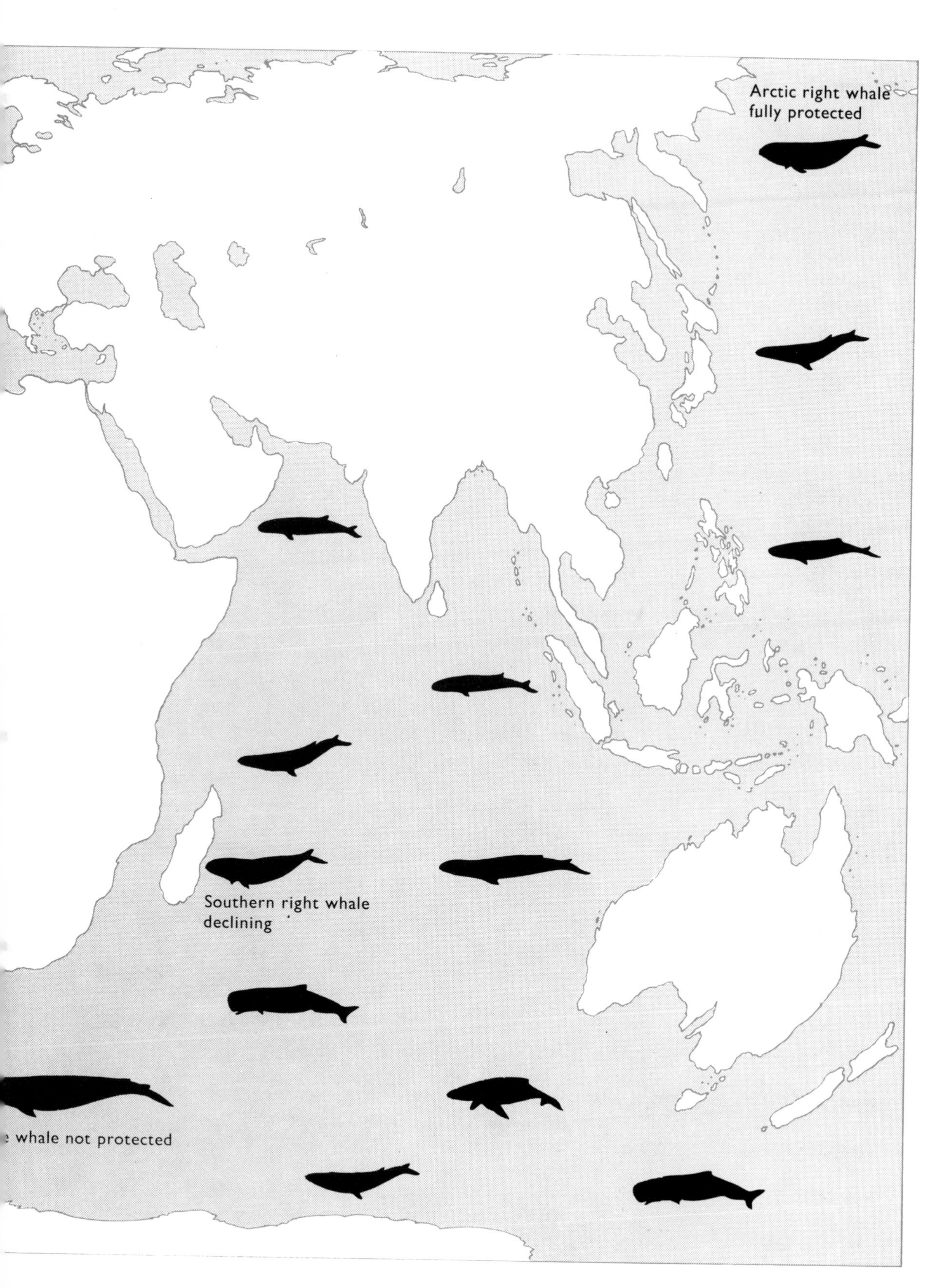
Arctic right whale
fully protected
Southern right whale
declining
whale not protected

The campaign

Despite all its good intentions, by itself the IWC was virtually powerless. It had a scientific committee to advise on quotas, but the advice was not difficult to ignore.

This judgement is not so harsh as it may sound. Quotas are not simply numbers plucked from the air. They must relate to the number of individuals that can be taken from a population without depleting it, the 'maximum sustainable yield' (msy). Calculating the msy is difficult for any species, and for whales it is almost impossible because the animals themselves are so difficult to study. They cannot be kept in captivity because they are much too large, and so scientists have to obtain their information from caught whales – caught by whalers, of course – and from sightings by ships, most of which will also be whalers. Whalers travel to places where they expect to find whales, and so numbers based on sightings or catches may exaggerate the size of the total population.

Whales can be tagged, by firing a metal tag into the skin, where it embeds itself without injuring the animal. A record of the whales sighted and tagged can be compared with the number of tags returned from whales caught by hunters, and the result can yield useful information about whale movements and the size of populations. It depends, however, on tags being found in the blubber and then returned to the scientists, and there is no way of knowing how many tags are simply lost.

It is also possible to relate the time spent hunting to the number of whales caught, since the fewer whales there are the longer it will take to find them. This calculation provides numbers, but they may be distorted by technological changes. There could be very few whales, but perhaps satellite observation could enable the whalers to find them in the time it took to sail to them by the shortest route.

The scientific advice could be challenged and was, by ardent whaling nations who could produce apparently plausible figures of their own. Nevertheless, no one could dispute the fact that during the 1950s and 60s catches of the more valuable species were declining.

In 1969 The Sierra Club, one of the leading US conservation organizations, split in two, when David Brower, a senior Sierra Club official, left to form a new organization called Friends of the Earth. In 1970 Friends of the Earth opened a British branch, and within a couple of years it had branches in many other countries. Almost from the day it was founded Friends of the Earth began its campaign to bring an end to commercial whaling.

Above *A Spanish whaler at work, photographed by Greenpeace in 1978.* ***Right*** Rainbow Warrior, *the Greenpeace converted trawler, sunk by French secret agents in Auckland Harbour in 1985. Greenpeace vessels tracked the whalers, recorded and publicized their activities, interfered as much as they could, and maintained ceaseless pressure to end whaling.*

Friends of the Earth differed from all other environmental and conservation groups. It did not register as a charity. Under British law a registered charity is entitled to tax relief but it is not permitted to engage in any political activity. It cannot even call for a change in legislation, for that is political. The tax advantage is great, but the political prohibition prevents serious campaigning. Friends of the Earth could campaign, and it did, in the most uninhibited fashion.

Rather than pursuing broad, general and somewhat vague objectives it selected 'targets' and concentrated on them exclusively. It compiled as much factual information as it could – and it was a great deal – and used this to influence politicians directly and to provide the background material for a popular publicity campaign that would win public

GREENPEACE

support for the actions it was asking of the politicians. An inflatable model of a whale was paraded through streets behind banners, advertisements were placed in newspapers, there were public meetings, demonstrations, and before long no one could claim to be ignorant of the fervent desire of at least some people to stop whaling.

In 1972, at the UN Conference on the Human Environment in Stockholm, a resolution was passed calling for a ten-year moratorium on whaling. Later this was confirmed by the UN General Assembly. It was not implemented because, as always, the IWC was split between non-whaling nations favouring the moratorium and whaling nations opposing it with an effective right of veto, but the campaign continued.

Greenpeace, a new environmentalist organization from Canada, intensified the campaign from 1977. It had a converted trawler, the *Rainbow Warrior*, and inflatables with which it sought to sabotage whaling directly – and with considerable courage on the part of the Greenpeace crew. Later it acquired other vessels, and worldwide publicity, but its most spectacular escapade occurred on 16 July 1983, when *Rainbow Warrior* crossed the Bering Strait from Alaska and landed a party of six anti-whaling protestors at Lorino, in the USSR. The protesters were arrested, and released after a few days, but photographs they managed to take out were rushed to Brighton, where the IWC was meeting. Greenpeace claimed that a plant in Lorino was processing meat from grey whales into feed for fur farms, which contravened international law. Rainbow Warrior was sunk in Auckland, NZ, harbour on 10 July 1985 while preparating to protest against French nuclear weapon testing in the Pacific.

Whaling did not end at once, but it was dealt a serious blow by an agreement among non-whaling nations to forbid the import of whale products. The ban was phased to allow substitutes to be developed, but it was fully in force by the late 1970s and closed many formerly lucrative markets.

In the end the pressure was irresistible and one by one the whaling nations agreed to a moratorium – effectively to an end to whaling. By 1985 Japan, the USSR and Norway were the only opponents to the moratorium, and by the summer of 1985 Japan had conceded, under pressure from the United States, which was required under its own laws to respond to continued Japanese whaling by curtailing, or even terminating, Japanese fishing in the US 200-mile Exclusive Economic Zone. The USSR was expected to follow, since almost all its whale products were exported to Japan. For the time being Norway continues alone, but its catch is very small.

The lessons

The anti-whaling campaign had achieved its objective, and had provided some lessons for all future campaigners. It had demonstrated that it is possible to alter policies not just of one government but of many governments by the use of well-rehearsed argument and widespread and vocal popular pressure, but it is not easy.

The issue must be clear, even if the arguments surrounding it are not. It is simple to demand an end to whaling, one particular activity that can be defined unambiguously. Not all issues fall into this category. The campaign against acid rain, for example, is much more vague because the causes of the problem are unclear and many activities contribute to it.

There must be a realistic alternative way for nations to behave. There were alternatives to whale products but in the case of the anti-nuclear power campaign, for example, the alternatives – fossil fuels or wind, wave, or tidal power – are either less obvious, or demonstrably more damaging than the campaign target, or both.

The campaigners must be prepared to work hard for a long time. People have been warning of the dangers of over-hunting whales since the 1930s, so it might be argued that it has taken half a century to bring the enterprise to an end.

It may also be true that a campaign will succeed only if the target is an activity that in any case is likely to decline because it is obsolescent. There is much inertia in society, so this is not so negative as it may seem. Whaling would have become increasingly costly and probably it would have collapsed economically, but the campaign brought it to an end sooner, and probably very much sooner, than economic forces would have done. Many pollution problems can be cured, or at least held within tolerable limits, by employing more sophisticated industrial technologies. Such technologies will be introduced one day, but campaigns to reduce pollution which make the older technologies more expensive because wastes from them must be retained or processed before discharge, accelerate their introduction. Campaigning can be a valuable political tool for accelerating desirable change, provided that change is eventually inevitable.

If this is so, it is unlikely that any campaign can succeed in stopping an activity that is expanding because its inertia is too great. It would be impossible to check the growth in the use of computers, for example – assuming someone thought it desirable to do so. There are too many of them already, they have too many uses, and we are committed to them irrevocably. The same may be true of recombinant DNA technologies – genetic engineering – and, for that matter, of nuclear power.

This leaves conservationists and environmentalists with many grey areas in which the future is not preordained – the exploitation of humid tropical ecosystems may be one – and where they may influence events. The success of the anti-whaling campaign shows that well-run campaigns can succeed, provided the target is chosen with great care, and that it may be possible, in principle, to compile a list of appropriate targets.

Demonstrators outside the Café Royal, London, where the International Whaling Commission was meeting in July 1979, calling for Britain to stop the import of sperm whale oil. Sperm oil imports were banned first in Britain, then in all EEC countries.

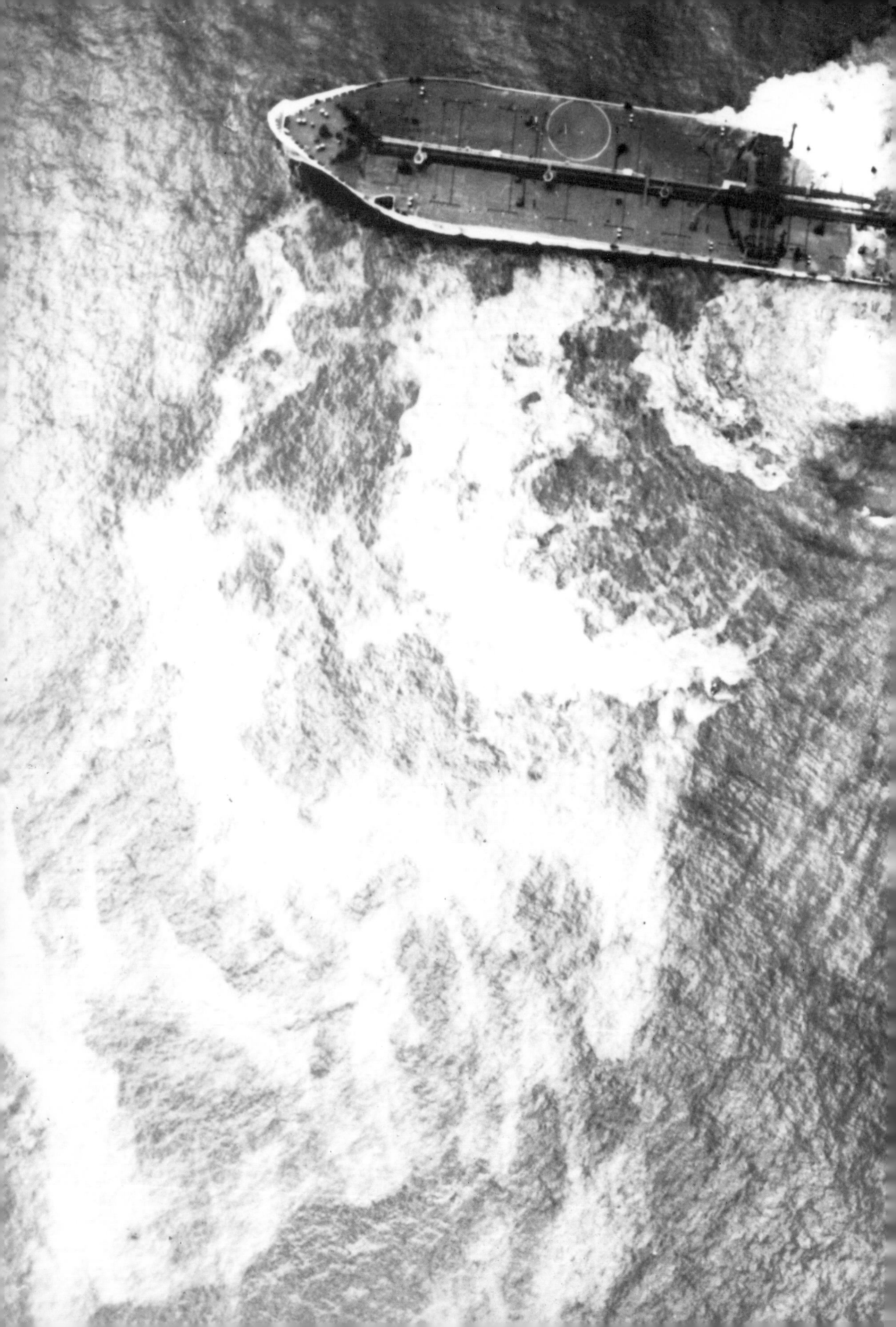

CHAPTER 10

The seas

The tropical rain forests extend over about eight per cent of the land surface of the Earth, and in places they are extremely rich ecologically. Area for area, however, they may not be the world's richest ecosystems and they are certainly not the strangest. For real ecological diversity and the truly exotic you must explore a coral reef, and the largest, most impressive, and richest coral reef in the world is the Great Barrier Reef, lying off the coast of Queensland, Australia. It extends from the Torres Strait, separating Australia and Papua-New Guinea in the north, to a point opposite Rockhampton, on the Tropic of Capricorn in the south, a distance of about 2,010 km (1,250 miles). In the north the Reef lies about 30 to 50 km (20 to 30 miles) from the coast, but in the south it is about 320 km (200 miles) offshore, all the way following fairly closely the 100-fathom (550 m) contour.

It is a 'barrier' reef because it lies offshore, parallel to the coast. If it extended from the shore it would be a 'fringing' reef, and if it were formed around the rim of a submarine volcano it would be an 'atoll'. Strictly speaking it is not one reef, but a chain of many reefs, more or less continuous in the north but more broken in the south. It is very vulnerable to pollution of the sea around it. That sea might be polluted in years to come because beneath it there is believed to be oil, and the Queensland government would like the area to be explored and, if oil is found, exploited.

In 1979 the Australian federal government passed legislation to establish the Barrier Reef Marine Park which would provide some protection, at least within the boundaries of the Park itself, but the Reef has another enemy, more difficult to control. There is a predator, the crown-of-thorns starfish (*Acanthaster planci*), which feeds on coral polyps and whose numbers have multiplied until in some places there are 15 of them to the square metre

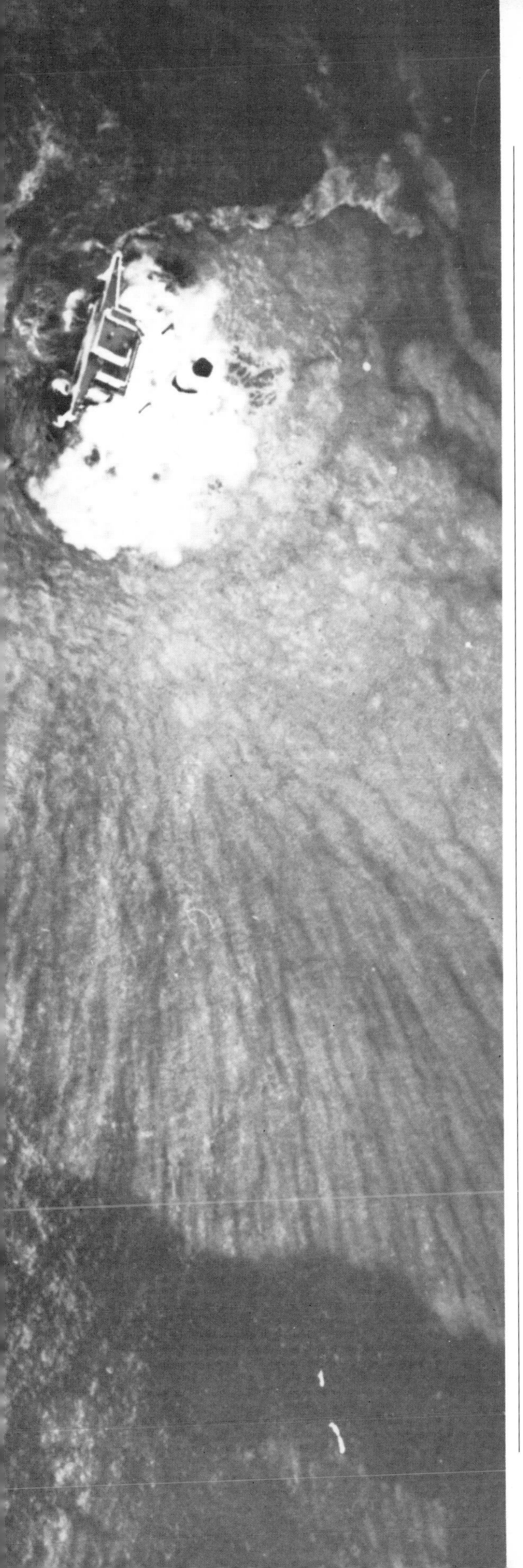

Amoco Cadiz, *the supertanker which ran aground on the Brittany Coast in March 1978 and later broke apart, spilling 16 billion barrels of crude oil.*

The crown-of-thorns starfish (Acanthaster planci), *which feeds on coral polyps, threatens Pacific reefs.*

(about 12.5 per square yard). They have their enemies, too, of course. The giant triton (*Charonia* species), a marine snail, and the painted shrimp (*Hymenocera* species) feed on the starfish and attempts have been made to increase their numbers, but a suspicion lingers on that it may have been some human interference that led to the proliferation of starfish in the first place.

So complex, yet so delicate

Corals are very small marine organisms related to sea-anemones and jellyfish. As juveniles they are free-swimming medusae – a jellyfish is a medusa that remains so throughout its life – but as adults they settle in one place, as polyps, secreting a tube of calcium carbonate for protection. They feed through the open end of the tube, using their tentacles to sweep the water for small items that are carried to the mouth, inside the tube. Many coral polyps live alone, but some are colonial and their colonies often comprise countless millions of individuals. It is these colonial polyps which build reefs.

Different parts of a coral reef are built by different species of corals. Some build the base and main framework, others fill in gaps and so hold the structure together, and the whole time the reef is under attack. Molluscs and sponges bore into it at the bottom, in areas abandoned by the living polyps, although the material excavated by the borers often buries and kills them, at the same time settling into holes and crevices and so consolidating the structure. Algae and other organisms may colonize abandoned surfaces. The number of coral species involved varies from ocean to ocean and reef to reef. The Great Barrier Reef, probably the richest, is made by more than 200 species.

The corals live in a symbiotic association with 'zooxanthellae' – algae which photosynthesize and which therefore need light. This means coral reefs can grow only where sufficient light for photosynthesis penetrates to the sea bed, and the water must be clear. Many other animals, including sea-anemones, sea-pens, sea-fans and sea-whips, which live on the surface of the reef, also depend on their relationships with zooxanthellae.

The water must be shallow and clear, and it must also be warm. Corals can grow only where the water temperature is 20°C to 28°C (68°F to 82°F).

The Great Barrier Reef is deep. Test bores have penetrated to more than 210 m (700 ft) without reaching bedrock. That is too deep for photosynthesis today, but at one time the sea was shallower because much of the planet's water was held in the ice sheets of the last glaciation. As the ice melted, the sea level rose, and the living corals built upwards to compensate. At other times the sea bed itself has subsided, and again the corals have grown back towards the light. That is how the Great Barrier Reef first started growing, in the Miocene Epoch, perhaps ten million years ago.

'Plankton' is the collective name given to all the tiny organisms that live near the surface of water, drifting with the current and wind and supplying food for many larger organisms. Clear, blue water contains very little plankton. Yet coral reefs support a vast assemblage of organisms for two reasons. They are built on the sea bed, and do not move freely in the water itself, so that many members of the community feed on the bottom and not from the water, and the algae photosynthesize, using water itself and dissolved carbon dioxide as their main source of food. The algae feed the population of animals that graze them, and the predators hunt the grazers. A coral reef is an ecosystem complete in itself and many of its members have adapted in remarkable ways to reef conditions. Many of the most popular marine aquarium fish are reef species, because they tend to be brightly coloured and marked – as camouflage against a brightly coloured background with areas of bright illumination and deep shade – and of curious shape.

Coral reefs grow extremely slowly and if they are damaged they recover extremely slowly – if at all. Coral makes a useful ingredient for cement, and reefs have been dynamited to obtain it, though this is now forbidden. Pollution of the water, by oil or by anything else that obscures light, will harm the zooxanthellae and everything that depends on them. Construction that causes changes in the flow of water around a reef may bring in pollutants or reduce the water temperature and this, too, is harmful. Many Caribbean reefs have been damaged as a result of human activities, and the Caribbean reefs were never so rich in species as the Indo-Pacific reefs. The Great Barrier is the richest of them all. Its destruction would be a major calamity.

The open sea

Some 70 per cent of the surface of the Earth is covered by water. It is easy to assume that all sea water is the same, all seas are the same. After all, it looks and tastes much the same no matter where you find it.

Biologically, however, the seas are different. The American continents separate the Atlantic and the Pacific. Most of the species which live in the middle and low latitudes of one cannot and do not visit the other. To do so would require them to pass, north or south, through water that is much colder than the water to which they are accustomed. The waters of the two oceans are kept apart by their own movements. It is only around the southern tip of Africa that the Agulhas Current carries water from the Indian Ocean – linked directly with the Pacific – into the South Atlantic.

The port of Balboa, on the Bay of Panama, at the Pacific entrance to the Canal, 26m below Gatun Lake.

The two oceans are not connected by the Panama Canal because the Canal rises by a series of locks to a maximum height of 26 m (85 ft) above sea level in Gatun Lake, and the central part of the Canal is fresh water. Should the present Canal ever be rebuilt at sea level – and the idea is proposed from time to time – the waters would be connected directly and this could cause ecological disturbance.

The rock forming the bed of the oceans is known as the 'oceanic crust' and the 'continental crust', which is geologically different, stands above it. Where the continental crust projects above the water there is dry land, but the continents themselves are partly submerged. The submerged region is called the continental shelf. Its area may extend more than 320 km (200 miles) from the visible coastline, and its margins slope more or less steeply toward the deep ocean.

Rivers, flowing from the continents into the oceans, carry with them particles of soil and quantities of organic matter. Fresh water is less dense than salt water and so tends to ride above it. The two mix only slowly, but little by little the river water loses its load of suspended matter, which sinks to the bed, but only on the shelf. This supplies nutrient to form the basis of the marine food chain, and so the continental shelf areas teem with life. Beyond the shelves, in the deep ocean, there is much less nutrient available, and consequently there are fewer living organisms.

A few currents carrying cold water from polar regions rise to the surface as they enter slightly warmer water. Water is at its most dense at 4°C (39°F). If that is the temperature of the surrounding water, water that has recently been in close contact with ice and is at a temperature of, say, 2°C (35.6°F) will be less dense and so will rise. As it rises it will bring with it nutrients scoured from the sea bed. Such currents, which include the Humboldt (Peru) Current which flows northward parallel to the west coast of South America and the Benguela Current,

The continents lie partly below the surface of the sea, the submerged areas constituting the continental shelves, shown here. At one time the continents were all joined, forming a southern supercontinent which split, most of it drifting north and splitting further. Because of the shape of the Earth, the continents broke into roughly triangular shapes pointing to the south.

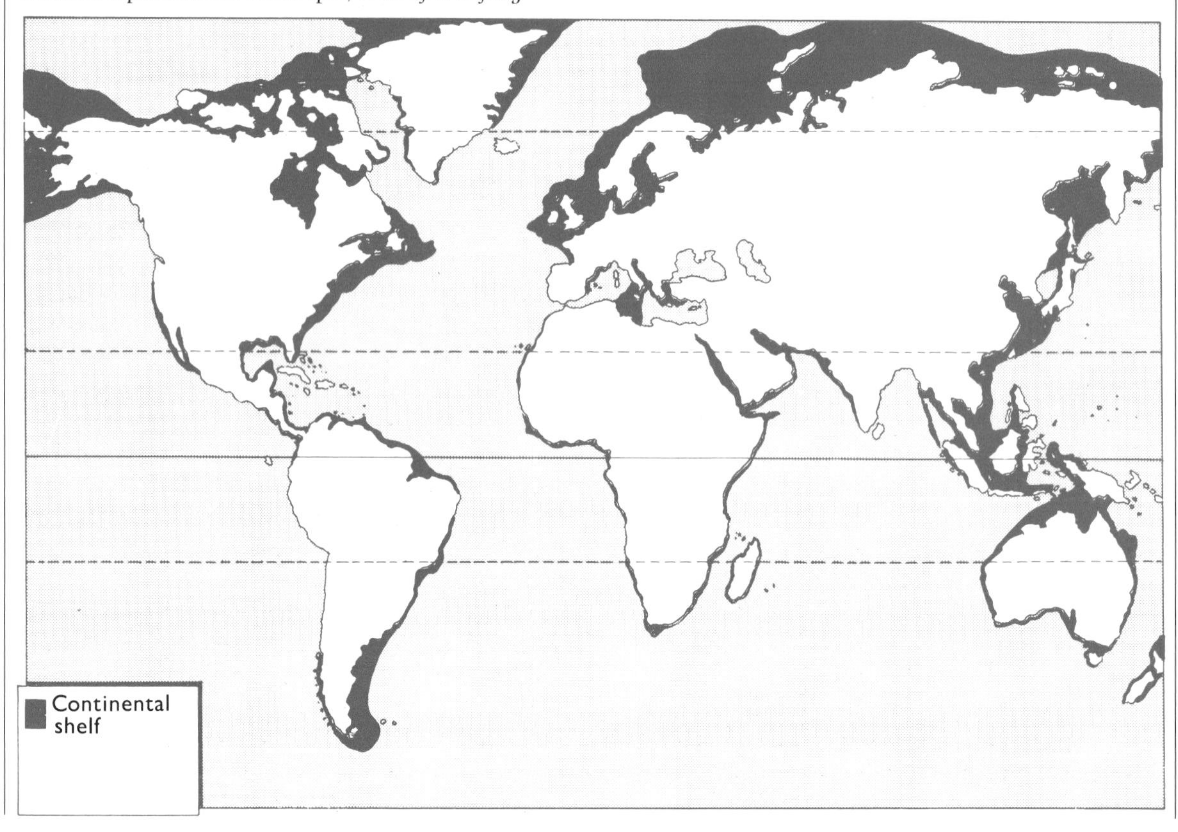

off the west coast of southern Africa, support many more organisms than does the ocean as a whole.

Commercial sea fisheries take place in these currents, but elsewhere they occur only over the continental shelves, and often very close inshore, because that is where the fish are to be found. It is one reason why nations guard their territorial waters so jealously.

Large areas of the deep ocean floor are desert. The nutrients in the sea support small photosynthesizing organisms – the 'phytoplankton' – which float near the surface. They feed small animals – 'zooplankton' – which in turn feed larger animals including most fish. The life is near the surface and the inhabitants of deep waters must subsist on its wastes and remains, which drift down to them constantly as a thin cloud.

A badly oiled guillemot. Its clogged feathers lose their insulation and buoyancy, and oil swallowed as the bird tries to clean itself may poison it.

Polluting the oceans

Material entering the oceans from the land may travel long distances, especially if it floats. Tree trunks are a well known hazard to racing yachts in mid-Atlantic. Plastic bottles, tattered bags, lumps of tar, and other flotsam have been found far out at sea. Because of the circulation of oceanic water they seem to be accumulating in the Sargasso Sea, the curious area of fairly still water in the central Atlantic where brown *Sargassum* seaweed drifts with the wind and supports a community of animals, some of them unique, and all of them more like those of a coastal area than of the mid-ocean.

If we discharge our own wastes into the oceans, clearly it makes a great deal of difference how we do it. Discharges close to shore will affect the most densely populated region. Whether or not they injure them depends on the nature of the wastes themselves. Often these are much less harmful than might be supposed, provided they are not washed ashore again to contaminate beaches used by humans.

Early in 1988 large blooms of algae released poisons that killed many fish in the North Sea. This incident aroused much public concern but the amount of phytoplankton had been increasing for some years, in the North Sea and in the Atlantic as well. The North Sea bloom may have been exacerbated by plant nutrients entering from the coast, but this is not certain and it cannot explain the increase in the central Atlantic. The plankton may be responding to increased carbon dioxide levels and slightly warmer sea-surface temperatures associated with the greenhouse effect.

The North Sea algal bloom was followed by the deaths of large numbers of seals and many people suspected a link between the seal deaths and pollution. It may be, for example, that exposure to PCBs (polychlorinated biphenyls) impaired their immune systems. The deaths themselves, however, were caused not by poisoning but by a viral infection similar to the one that causes rinderpest in cattle.

We should not be complacent, for while such pollutants as radionuclides and oil, which cause most public concern, are of no great biological importance others may be, and in ways of which we are only dimly aware. Inshore waters support the most abundant marine life, but they are also the location of chemical processes by which organisms cycle essential elements back to the land. The muds on the sea beds and in estuaries play a vital part in this. If we interfere with this chemistry the consequences might be grave.

Unwise coastal development might cause even more damage than the discharge of effluents. The draining of coastal wetlands – swamps and marshes of brackish water – is popular, because it provides dry land on which to build hotels and other tourist attractions. Yet the wetlands are natural chemical complexes which also provide habitat for many rare species of plants and animals. Today they are the subject of growing international concern and inter-governmental bodies such as the Council of Europe have urged their protection. In March 1983, at Cartagenas de Indias, Colombia, 27 nations signed the Cartagena Convention for the protection of the marine environment of the Caribbean, and the United Nations Environment Programme (UNEP) began to seek funding from governments for the $1.5 million Caribbean Action Plan which it sponsored and which is called for by the Convention. This will begin with the protection of coastal waters in general and mangrove swamps in particular. The protection is designed to preserve the species, but if it succeeds it will also safeguard the chemistry.

Polluting the deep ocean

Inshore pollution is much more serious than the discharge of wastes into the deep oceans, and even pollutants discharged from coasts are rendered harmless once they move beyond the continental shelves. The reason is simple. They are diluted in such a large volume of water that even if they are not rendered harmless by undergoing chemical reactions with ordinary constituents of the water they will exist only as isolated molecules. With the smaller seas that are linked with it directly, the Atlantic Ocean has a surface area of more than 106 million square kilometres (41 million square miles) and a volume of about 400 million cubic kilometres (100 million cubic miles). The Pacific has an area of 330 million square kilometres (64 million square miles) – and a volume of about 1,400 million cubic kilometres (170 million cubic miles). It is a large amount of water in which to disperse a few tonnes or even thousands of tonnes of garbage.

We need not worry about the disposal of sealed containers of wastes in the deep ocean – not even those containing soiled protective clothing, laboratory ware and other sources of low-level radioactive wastes. Such low levels of radioactivity would be harmless to marine organisms even if the containers were to leak, which is improbable. The containers sink to the ocean floor and lie in water that moves

The Mediterranean Basin, showing the complex of small depressions which comprise it, and sources of pollution feeding into it.

only slowly and moves to the surface even more slowly. If a container were to leak, its contents would have ceased to be radioactive by the time they could contaminate fish swimming near the surface, or any human could come into contact with them.

The closed seas

There are international treaties regulating the dumping of wastes at sea and in 1973 the London Dumping Convention called for a two-year moratorium on the dumping of radioactive waste at sea. Until then the United Kingdom had been one of the few countries still disposing of wastes in this way, dropping them in the Atlantic, about 500 km (300 miles) off Land's End, and beyond the edge of the continental shelf.

It is hardly appropriate to forbid the convenient disposal of wastes in properly constructed and sealed containers in deep water when the alternative is much more difficult disposal on land, but observation of such regulations may be a price worth paying if it means the true aim of the regulations is achieved. They are intended to deal with the real problem of shallow seas, and especially those that are more or less land-locked and so isolated from the oceans and the diluting effect of a vast volume of water. The North Sea, the Baltic, the Persian Gulf, the Red Sea and the Caribbean are examples of such seas, but the classic example, and perhaps the most extreme one, is the Mediterranean.

The Mediterranean is not one sea, but several linked seas, for it includes the Tyrrhenian, Adriatic and Aegean Seas and it is linked, through the Dardanelles and Sea of Marmara, with the Black Sea, which is linked in turn by Kerch Strait to the Sea of Azov. A submarine ridge running from western Sicily to Cape Bon in Tunisia divides it into two basins and another ridge, from Crete to Cyrenaica in Libya, further divides the eastern basin.

The Mediterranean comprises a set of submarine basins separated by more elevated areas. Water enters from rainfall, from rivers, and through the Gibraltar Strait, and leaves mainly by evaporation, although some water returns to the Atlantic.

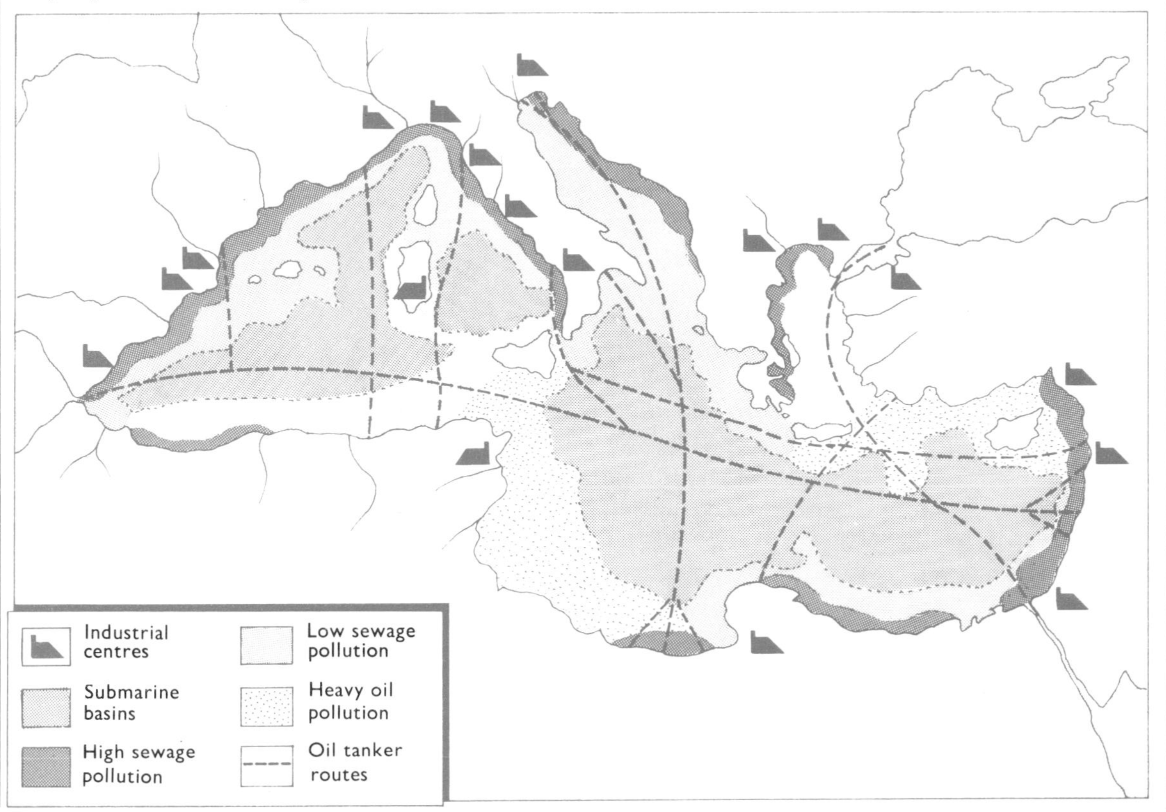

Above Discharge outlet and box for dumping garbage over the side of a ship.

Right Sewage outfall pipe, discharging untreated waste from the British coast directly into the sea.

The whole complex is linked to the Atlantic at only one point, the Strait of Gibraltar. Water enters by surface currents from the Atlantic, from rainfall, and from rivers, but most of it is lost by evaporation. In winter, when the addition of water by rainfall exceeds evaporation, surplus water leaves through the Strait of Gibraltar in currents moving close to the sea bed. Bottom currents also feed the Black Sea, and water leaves the Black Sea close to the surface. Because much more water enters the complex than is returned to the Atlantic the water is exchanged for ocean water only very slowly and any substance discharged into the Mediterranean tends to remain there. There is ample opportunity for such discharges. Taken together the seas receive the waters of the Ebro, Rhône, Po, Danube, Dnieper, Don and Nile, and of many smaller rivers. The Mediterranean is also used by shipping proceeding to and from the Suez Canal – and the oil terminals in the Gulf. Tankers travelling empty to collect oil cargoes often wash out their tanks in the Mediterranean, out of sight of land because this practice is forbidden, and about one-quarter of all the world's oil pollution occurs in the Mediterranean.

The reduction in pollution requires the cooperation of the countries bordering the seas. They must reduce discharges of pollutants into the rivers that feed the seas, and they must collaborate to control the activities of ships passing through their area. Since the objective is so obviously desirable, it all sounds simple enough until you draw up a list of the countries whose agreement is sought. These are: Spain, France, Italy, Yugoslavia, Albania, Greece, Malta, Bulgaria, Romania, the Moldavian SSR, the Ukrainian SSR, the Russian SSR, the Georgian SSR, Turkey, Syria, Cyprus, Lebanon, Israel, Egypt, Libya, Tunisia, Algeria, and Morocco, with the addition of the United Kingdom if Gibraltar is not included with Spain. Broadly, there are developed, industrialized countries on the northern side, socialist countries to the east, and the developing countries of North Africa to the south.

It might seem a hopeless task to persuade such a disparate group of nations, whose interests and

ideologies conflict at almost every point, to agree about anything. Nevertheless UNEP arranged conferences, drew up and persuaded governments to pay for a Mediterranean Action Plan whose fourth protocol was signed in 1982 by 18 of the states bordering the Sea, and a number of treaties, the most recent of which, the 1980 Athens Treaty on Land-Based Sources of Pollution, came into force in 1983 when 6 of the 16 signatory countries ratified it.

This formed part of the UNEP Regional Seas Programme, each stage of which deals with a particular area and when a plan to protect it has been agreed hands over the responsibility to the governments involved. Such plans now exist and either have been or soon will be implemented for the Gulf of Guinea and other waters off West and Central

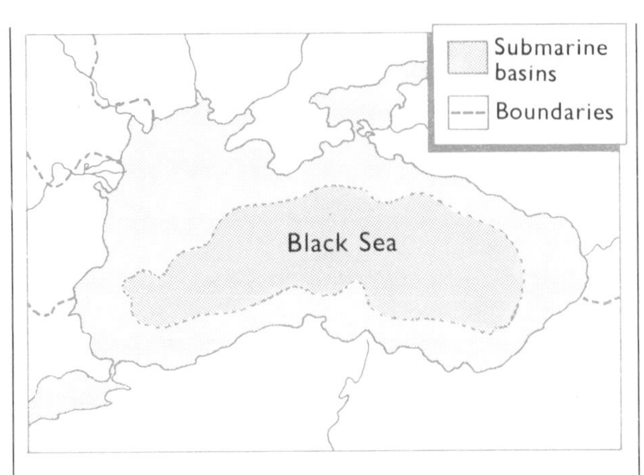

The Black Sea, formed from a single basin, forms the eastern part of the Mediterranean Basin complex, and is fed by rivers, rain, and from the adjacent sea.

The 265,000 ton supertanker British Ranger. *Ships of this size can cause extensive pollution simply by cleaning their tanks illegally but cheaply, at sea. Recession and an oil glut have made most of them uneconomic and today they lie idle.*

Africa, the Persian Gulf, the Red Sea, and various waters in East Africa, East Asia and the Pacific, as well as the Caribbean and Mediterranean.

Protection of the Baltic is covered by the Helsinki Convention for the Protection of the Marine Environment and the North Sea is protected by other agreements, including the 1974 Paris Convention for the Prevention of Pollution from Land-Based Sources, the 1972 Oslo Convention, and several agreements drawn up by the Intergovernmental Maritime Consultative Organization (IMCO). UNEP attention is now concentrated on areas of the Indian Ocean and Pacific.

Fish and metals

Until very recently our interest in the sea has centred on what we can take from it rather than what we can dump in it.

Commercial sea fishing provides us with nutritious food, rich in protein and unsaturated fats, and it provides those engaged in it with employment. Before railway lines were opened and efficient means were available for freezing, marine fish were eaten mainly in coastal areas.

Today people expect to be able to buy fresh fish – albeit frozen – no matter where they live. This has increased demand, brought prosperity to fishing communities, and led to the development of much more effective fishing technology.

The raising of marine fish from eggs to market size in captivity – literally fish husbandry – is possible for many species but little practised because it is more expensive than hunting, and so to this day fish are hunted. They are hunted, however, by two quite distinct methods. The traditional method involves small vessels, up to about 18 m (60 ft) long but often much smaller. Each has a crew of up to about six, who are paid a share of the value of the catch, they

Areas covered by the UNEP regional seas programme.
1 Mediterranean
2 Kuwait
3 Caribbean
4 West and Central Africa
5 East Africa
6 East Asia
7 Red Sea and Gulf of Aden
8 South-west Pacific
9 South-east Pacific

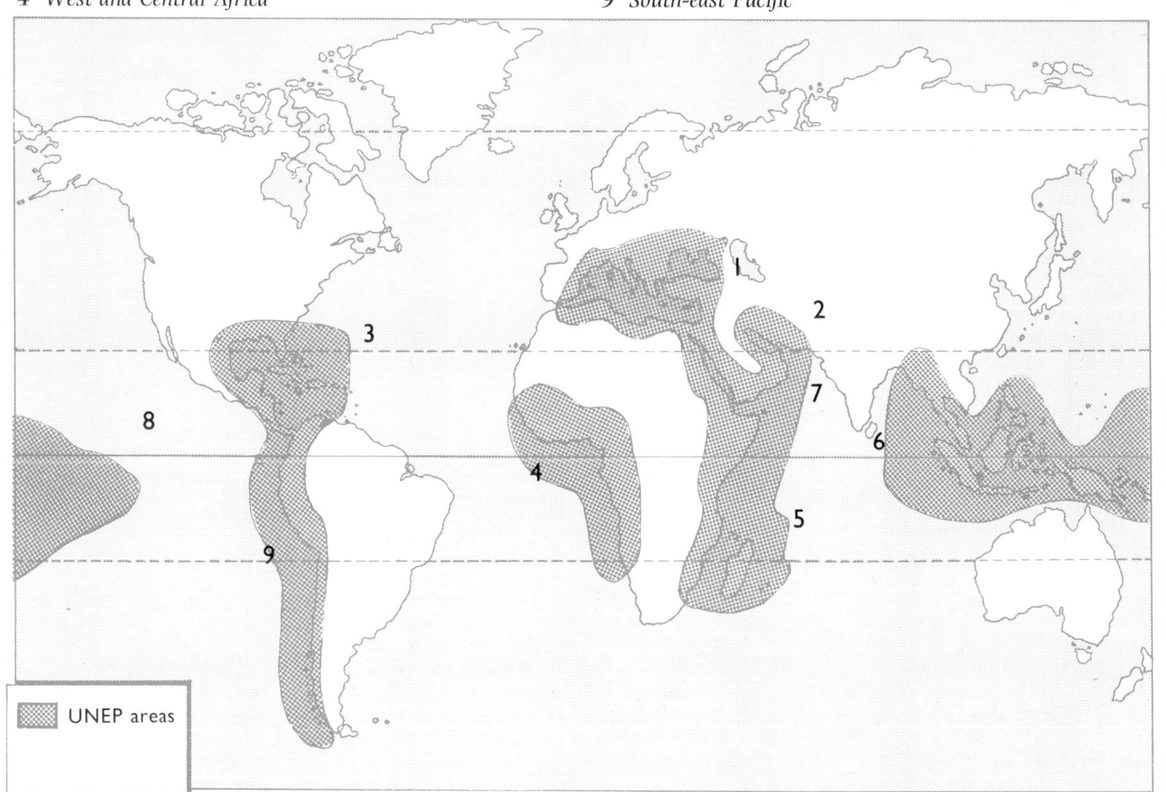

Left Salmon farming, Washington State. Harvested fish, stored in an ice tub, are being loaded into barrels on their way to market. ***Top*** *A small inshore trawler sailing from a British fishing port. Inshore fleets provide enough employment to support whole village communities and do not harm fish stocks.* ***Above*** *A Spanish fisherman with a bluefin tuna or tunny* (Thunnus thynnus). *They can be nearly 4m long and weigh 700 kg.*

work close to shore and often within sight of land, and when weather conditions are bad they do not sail. Their rivals work in much larger vessels and in all weathers. Both types of vessel use the same basic techniques of netting, but the larger vessels do so on a much larger scale. They can catch an entire shoal in one haul of their net.

The essential difference arises because of the costs of the two kinds of operation. The small vessel is relatively cheap to buy and operate, but there are many of them and so a fleet of them can support an entire village community. The large vessel is much more efficient, and much more expensive. It uses more fuel, but its economics are dominated by the cost of servicing the loan usually involved in buying it and in high insurance premiums. It employs a larger crew, but they must be paid wages, not a share in the catch, and they must be paid whether they go to sea or stay in port. Consequently, the larger vessels cannot afford not to fish. They have a greater range and so can travel the world seeking fish wherever they can find them. Much of their catch is sold for direct human consumption, but a high proportion is also sold to make fishmeal, for feeding farm livestock or as fertilizer.

These distant water fleets, with their large expensive vessels, cannot avoid competing with the inshore fleets, but they do not compete on equal terms, and protecting the livelihoods of the inshoremen has proved difficult everywhere in the world. Quotas have been imposed to restrict the quantity of fish of a particular species that may be caught during a season or year, but a few large vessels can catch in a week or two fish it would have taken the inshore fleet months to catch, so the quota can be fished out rapidly, leaving the local fishermen with nothing. Markets, developed by the inshoremen, can be flooded by catches from the distant water fleets, so that prices plummet.

This conflict often translates itself into fears of overfishing, and over-fishing can deplete local stocks. This is a rare event, however. Fish populations fluctuate naturally and are very susceptible to quite minor and natural changes in the chemical composition of their waters. The anchovies that formed the basis of a huge Peruvian fishing industry – for fishmeal, not for human consumption – disappeared in the early 1970s, but not because of overfishing. The fish lived in the cold Humboldt (Peru) Current, and the Current moved as part of a general change in conditions in the Pacific, known as the 'El Niño Effect' but still not clearly understood. North Sea herring also disappeared, years earlier, but this, too, was a natural and not a man-induced phenomenon.

Fish are caught over the continental shelves, and the seabed of the shelves is also rich in mineral and fuel resources. Offshore oil and gas production is now familiar, but there are also coal and metal ores, and 'manganese' nodules. These are roughly spherical masses of metals, many of them including manganese but with other metals as well, that lie on the seabed awaiting collection by those with adequate technology.

The Law of the Sea

Early in the 1970s the United Nations began a series of conferences, each of which was called the United Nations Conference on the Law of the Sea, or UNCLOS. The idea was to draw up a comprehensive treaty that would regulate all human activity in the oceans, and more especially beneath the surface of the oceans. This, it was felt, could be used to conserve fish stocks and so protect the livelihoods of inshore fishermen while controlling the exploitation of seabed minerals of all kinds.

The aim was noble, but progress was slow, mainly because of the insistence of countries which have no coastline nor a share in the management of marine resources. The way proposed to achieve this, at least in respect of seabed minerals, was to require all mineral exploration and exploitation to be licensed by a central Seabed Authority, based in Jamaica. A condition of the licence would be that a share of any profits was paid to the Authority, eventually for the benefit of Third World countries, and that the Authority be allowed unrestricted access to all technological details of the operations. This would permit the UN to distribute technological information freely to all those who might need it.

The conditions proved unacceptable to several of the countries – including the USA and United Kingdom – most capable of developing, and therefore most likely to wish to develop, seabed industries. In particular the Americans were not prepared to agree to parting with technological information because, they maintained, the technology was often leased, and not owned, by the companies using it. It might therefore be illegal for them to pass it on, and so they could not be expected to agree to that condition of the licence.

Stadrill, *a semi-submersible self-propelled drilling platform undergoing trials in the Firth of Forth before moving to the Cormorant field in the North Sea. At present oil and gas are the most sought-after products of the sea bed.*

The one concept that emerged from the Conferences and which was seized rapidly, and eventually by almost all countries, was that of the 'Exclusive Economic Zone' (EEZ). It allows coastal nations to control all commercial activity, theoretically over the part of the continental shelf which extends from their coasts straight out to sea as far as the 150 m (500 ft) depth contour, which marks the beginning of the slope toward the deep ocean. In practice, since not all coasts open on to continental shelf, the limit is set at 320 km (200 miles) from the shore and where two countries face one another across a narrow sea area and so have to share an EEZ, a median line is agreed privately between them.

The Law of the Sea Treaty should have been signed by December 1984 but the deadline came and went and the USA and UK still refused. Perhaps, one day, it or some version of it will be agreed and will come into force.

The protection of the oceans and seas from pollution and the rational exploitation of their resources concerns the coastal areas and 'regional' seas greatly, the continental shelf areas only slightly less, but the vast expanses of the deep oceans hardly at all. If there are resources beneath their beds they are quite inaccessible to us with our present technology, and will always be difficult and costly to work, while the volume of water in the open oceans is more than adequate to dilute any waste we may take the trouble of travelling so far out to sea to dump in them. The problems of the continental shelves, coastal areas, and regional seas are recognized and in one way or another appropriate action to safeguard them is being taken.

The deep oceans are remote, but this does not mean more general protection of the marine environment is remote, and as individuals or members of conservation or environmental groups we are not helpless. If you wish to help there are a few guidelines that may be useful. Waste disposal in the deep oceans is harmless provided the wastes are sealed in containers that sink to the bottom. Developments close inshore, on the other hand, may be far from harmless. The draining of coastal swamps and marshes and interference with river estuaries may disturb many organisms, including the young of some species of commercially important fish, and may disrupt vital chemical processes. Marinas may lead to damage, except where they are built in existing harbours. Such developments as these should be regarded with extreme caution, and opposed in most cases. Proposals to establish marine nature reserves in order to protect especially valuable or interesting areas should be supported.

UK fisheries limits are defined as comprising the sea area lying within 200 nautical miles of the coast, except where median lines have been agreed with other governments in respect of seas less than 400 nautical miles wide. Within this sea area all fishing is subject to UK regulation.

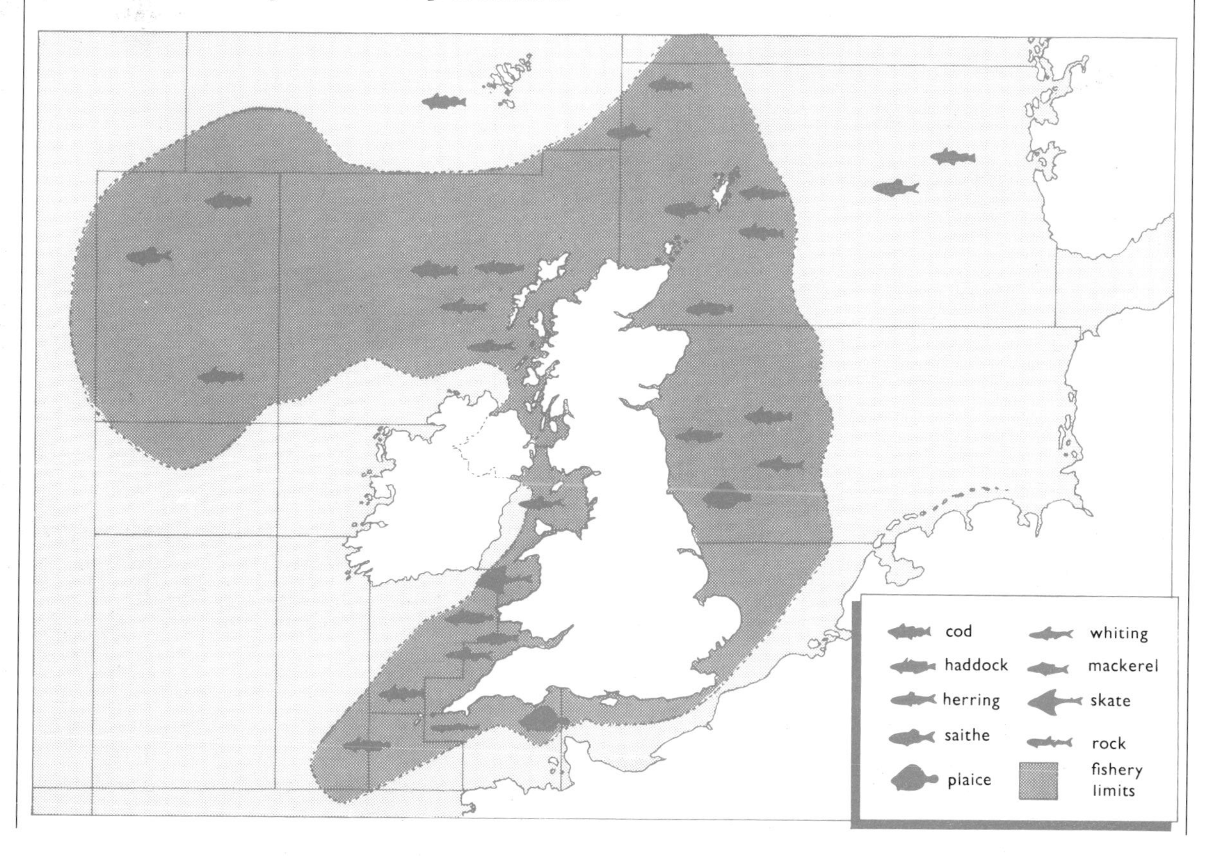

Left *Lady Elliot Island, surrounded by a lagoon bordered by a barrier reef, is part of the Great Barrier Reef.* ***Below*** *Sea anemones are carnivorous or suspension-feeding animals closely related to coral polyps (both belong to the class Anthozoa, 'flower-animals'), and often live on reefs, but they are usually solitary and have no mineral skeleton.* ***Right*** *Part of the Great Barrier Reef, possibly the world's most diverse ecosystem, but now threatened by predators and pollution.*

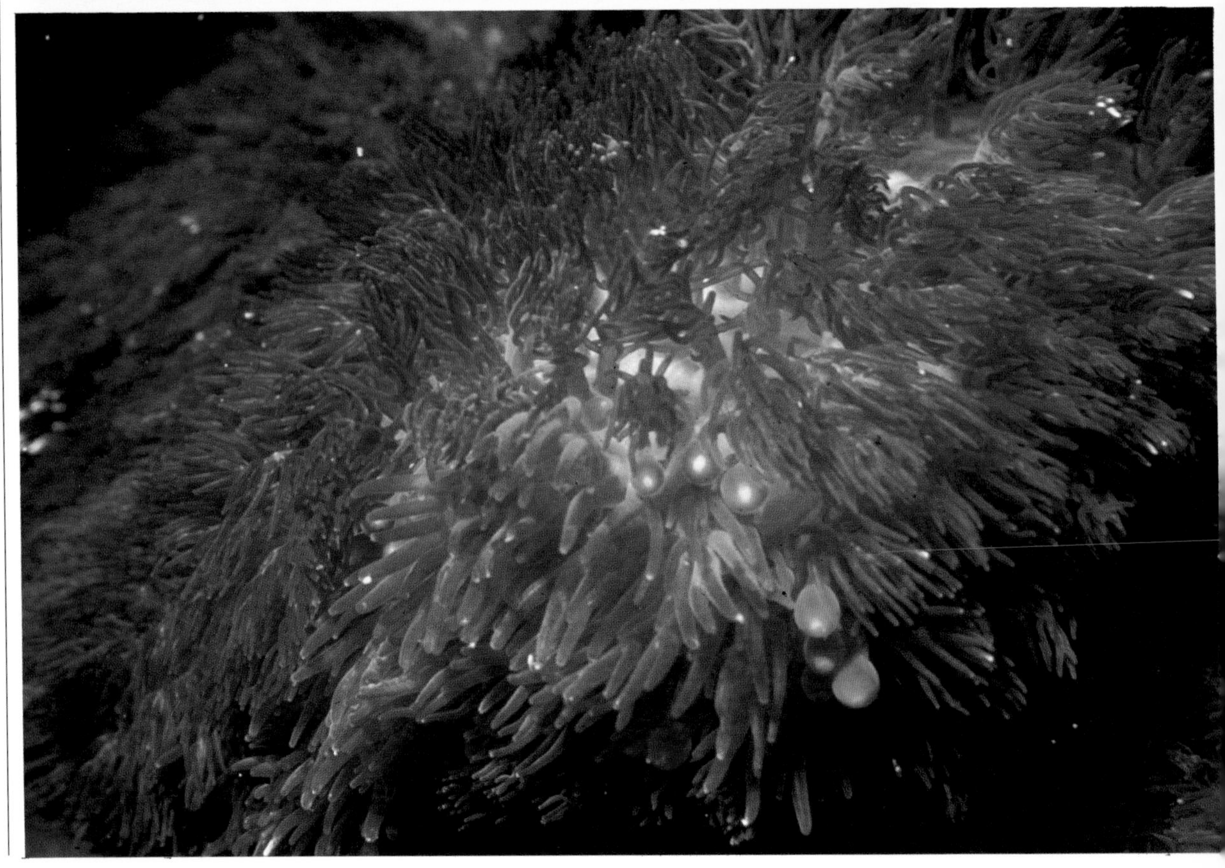

CHAPTER 11

People, hunger and poverty

Most of the environmental problems facing us are novel but famine is not like them. It has existed for thousands of years, perhaps for as long as there have been humans and certainly for as long as humans have produced their food by farming. It is the human problem about which we should know most, and the one we should have solved long ago.

Famine is not new, but it has changed. A little more than a century ago there was severe famine in Ireland. Earlier this century there was famine in parts of the United States. Today it is unthinkable that there should be famine in North America, Europe or Japan unless some major catastrophe altered radically the economic situation of those regions. Modern famines do not affect the citizens of rich industrial countries. They happen only in poor countries. Why?

The most common popular explanation can be summed up in one word: overpopulation. There are too many mouths for the farmers of the world to feed. It follows from this that famine can be eradicated from the world only by the removal of some of the people. We may prevent people from being born, or we may watch with such equanimity as we can muster while the surplus die. In effect we may eliminate poverty by literally eliminating the poor. The explanation is uncomfortable. It is also dishonest, and plain wrong.

One comparison should be enough to illustrate the dishonesty. Along with most countries in Africa, Ghana has been urged to reduce its rate of population growth, to prevent the size of its population from increasing. The area of Ghana is about the same as that of Great Britain, and most of the country is low-

Camp at Korem, northern Ethiopia, where in early 1985 the Save the Children Fund was feeding 12,000 children a day. Emergency supplies are essential when disaster strikes; but more complex and far-reaching measures are needed for chronic poverty and hunger.

lying, good farming land. Much of Britain is marginal upland, of little use for food production. The population of Ghana is around 12 million. The population of Britain is around 60 million. The real difference between the two countries is that Ghana is much poorer than Britain.

The dishonesty arises from errors in the overpopulation theory itself. This was first proposed in 1798 in a book called *Essay on the Principles of Population*, written by the Rev. Thomas Robert Malthus and at first published anonymously, but with a much revised second edition published in 1803 under Malthus's name.

The theory became highly influential, probably because it is very simple indeed. Malthus argued that parents are capable of producing offspring in numbers large enough for the population as a whole to increase. Two parents may have four children, for example, and each child may grow up, marry, and the married couple have four more children, so that numbers double in each generation: 2, 4, 8, 16, 32, 64, 128, 256. This is growth by compound interest, or geometric or exponential growth. The resources to feed people, Malthus maintained, do not increase in this way. If you increase food output, say by ploughing a few more acres of land, this reduces rather than increases the ease with which further gains can be achieved. So, at best, food production – and the same is true for the provision of all other physical necessities – increases only by simple interest, or arithmetically: 1, 2, 3, 4, 5, 6, 7, 8. As time goes on, therefore, the difference between demand and supply widens, and this must lead, inevitably, to the death of those whose needs cannot be met. Any attempt to increase the availability of resources will serve merely to encourage further population growth, because – and this is the 'law' he propounded – population will always tend to increase to the very limit of the resources available to sustain it.

Malthus was writing about England around 1800, and despite his apparently self-evident arithmetic he made three mistakes. The first was that the population was growing much more slowly than he supposed. The second was that if he was correct, population should increase fastest among those people with the best access to resources – the rich. It did not then, any more than it does now. It was the poor who had large families. The third mistake was his underestimation of the effects of changes in farming – the ease with which necessities may be provided. Even when he wrote, living standards were improving, as new farming methods became established in Britain and crop yields began to rise.

More rational methods of dealing with the problem of world hunger are based on reproducing in poor countries the kind of changes that took place a century or more ago in what are now the rich countries.

Famines, old and new

Famine is caused by a partial failure in the supply of food. The harvest may fail because of crop disease, pest infestations or bad weather, or it may fail because it has to be abandoned. Warring armies may prevent farmers from working their land, or may even drive them from their homes, to wander with their families as refugees seeking whatever help they can find while their abandoned crops rot in the field, so there is less food for everyone.

The shortage leads to a rise in food prices, because prices are determined by supply and demand. Those who can afford it have to pay more. Those who cannot afford the new prices go short, and if the food supply fails to improve, stocks are reduced, prices rise still further, and the number of people who cannot afford to eat increases. In rural areas, where generally wages are much lower than they are in the cities, people may be forced by hunger to eat the grain they had been saving to sow next season. This means they will be short of seed in the following year, so there is a likelihood that the famine will continue at least for two years and possibly for longer.

At one time the effects of a famine would be confined to the region in which it occurred and as many people as could would leave their homes and go in search of better conditions elsewhere. In the nineteenth-century Irish Famine, for example, some two million people died, but another two million emigrated, mainly to the United States. Throughout history poor people have always moved when conditions become too harsh.

Sunflower harvest in Kenya. The seeds are highly nutritious, but drought has devastated the crop. Farming employs one-fifth of the labour force but only 10% of the land is in permanent cultivation or pasture.

A famine could not occur in Britain now because the British people would import the extra food they needed, buying it wherever they could find it and paying whatever they had to pay, and there is no mystery about where they would find food. The United States grows far more food than its own people can eat and exports the surplus. The European Community, Australia, and New Zealand also export large amounts. There is no shortage of food – at a price. In the last half century or so food has become a major item in world trade. That is the first change to affect the pattern of famine.

When food is traded its price depends on the amount being sold and the demand for it. If the demand increases, so does the price. If Britain, say, suffered a serious harvest failure it would need to buy a large amount of food, and this would cause the price of food to rise in world markets. If another country also wished to import food it would have to pay the new, higher price, but if that country were very poor and had insufficient money its people would have to go hungry. During this century we have learned to avoid famine by exporting it. It is still the poor who suffer, but on a global rather than a local or national scale.

Why, then, do people not do what they have always done, and move to places where there is more food? Because we will not allow it. When the famine was local, better conditions might be found just a few miles, or even a few hundred miles away, and people could travel that far. Today it might not help, because the food might be thousands of miles away, and out of reach. Even if they could travel so far, the poor people would face political obstacles. Britain, to continue with the same example, will not allow poor people from other countries to enter. Neither will most other rich countries. They fear that unrestricted immigration would increase competition for jobs, housing, services of all kinds, and so reduce the living standards of their own people, and probably they are right. If we share resources fairly, this must mean that the poor have a little more, and the rich have a little less. We forbid the traditional remedy.

Cash crops

It is not the only remedy, or even the best one. It would be much better if the poor were to grow more food themselves, or produce goods they could sell to pay for the import of the food they needed.

Store in Wiltshire, filled with surplus grain resulting from EEC efforts to attain self-sufficiency. It may be released following a poor harvest, or eventually destroyed.

What shall they produce for export? They are poor, after all, precisely because they lack modern industries to produce goods with a high value in world markets, so in many cases they have little choice but to export agricultural products – cash crops – grown for sale rather than for local use. Between 1967 and 1972, cotton production in Mali quadrupled and peanut production almost doubled, and Upper Volta did even better. In 1960 it grew 2,000 tons of cotton; in 1984 it grew 75,000 tons. In both these countries the production of food for local consumption has been declining year by year. The reason is simple enough. Cotton and peanuts sell for a much higher price to rich foreigners than local people can pay for the sorghum or millet that can be grown on the same land, so farmers grow cash crops in preference to food crops.

The sale of high-value export crops could have paid for the import of fuel and machines which could have been the basis for improved farming methods and manufacturing industry. In some parts of the world that has happened and people are more prosperous, and better fed. In others, including the poorest countries of Africa, it went wrong. Oil prices rose, so fuel imports became more expensive. The cash crops had to be exported as raw materials rather than as finished goods, partly because the rich importing countries imposed taxes on imports that competed with their own industries but not on raw materials for those industries, and partly because commodity prices were unstable. A fall in the world price of cotton could bring disaster to a country dependent on cotton exports. Money was borrowed to pay for investment, but it was soon gone, and then exports did little more than pay the interest on the loans. In some cases they did not even do that and the debts went on growing.

Advancing deserts

The problem is most acute in those African countries that lie along the southern edge of the Sahara Desert – the Sahel zone, which includes all or part of Mauritania, Mali, Upper Volta, Niger, Nigeria, Chad, Sudan, Senegal, and Ethiopia – but it extends further, into East Africa and parts of south-western Africa. The Sahara is spreading south and the southern deserts are spreading north.

The spread of the desert is caused in the first place by drought in the countries that border it, which in turn is caused by a change in the climate. Close to the equator the ground is warmed by strong sunshine,

Drought in Niger drives people to gather leaves for food after conventional crops have failed. The leaves are edible, although hard and bitter, but stripping the trees may kill them, leading to further soil deterioration.

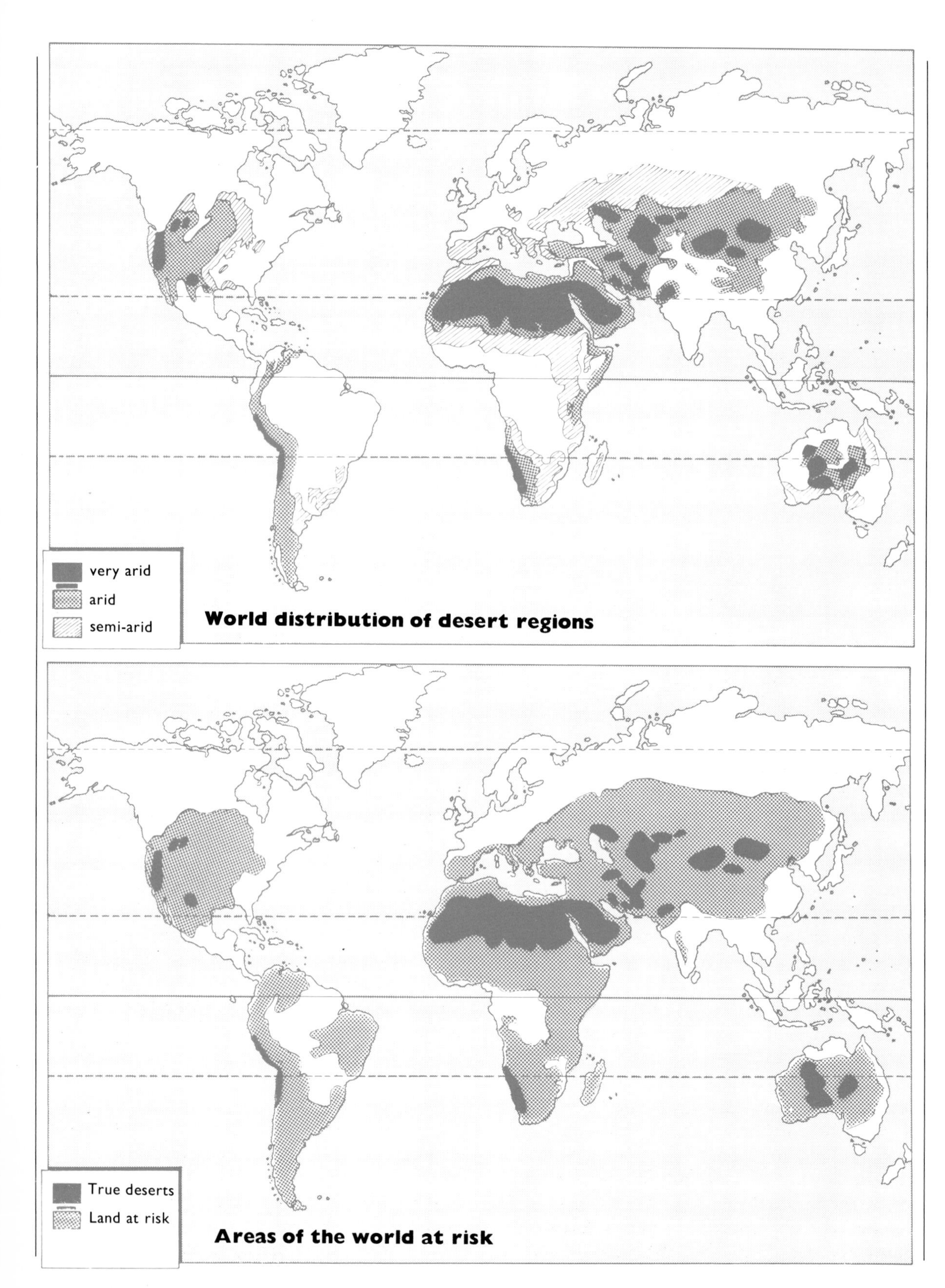
very arid
arid
semi-arid
World distribution of desert regions
True deserts
Land at risk
Areas of the world at risk

Nomads, with their camels, asses, sheep, goats and cattle, around a well in western Sudan. Mixed herds use the range more efficiently than would just one species. The people know their environment better than most outside advisers.

warms the air in contact with it, water evaporates, and the warm, moist air rises. As it rises it cools, its water vapour condenses to form clouds and rain, and the cooled air, which is now dry, descends in the subtropics. There is a boundary, the Inter-Tropical Convergence Zone (ITCZ), between the circulation of tropical air and that of the air in higher latitudes.

The Sun appears to move north and south of the equator with the seasons, and as it does so the ITCZ moves as well, so that equatorial rains fall in higher latitudes. This brings summer rain to the countries south of the Sahara, but not to the Sahara itself because the ITCZ does not move so far. Instead of rain the desert has the dry weather that has been shifted from its southern margin, and its winter weather also moves north, to bring a warm, dry summer to the Mediterranean region. Since the 1960s, climatic changes have had the effect of compressing the equatorial belt, so that the ITCZ can no longer be relied upon to move as far from the equator as it did formerly. In the countries of the Sahel zone, the seasonal rains fail, and go on failing. They are failing in the same way and for the same reason along the northern edge of the southern-hemisphere deserts.

The rains were never completely reliable, and droughts have occurred many times before, but today they are compounded by further problems, which are the second reason the Sahara is spreading.

The land on the edge of the desert is unsuitable for most kinds of farming. Much of it is poor grassland, inhabited by semi-nomadic pastoral peoples who drive, or more often follow, their herds and flocks in pursuit of such pasture as they can find for their animals. It is a precarious existence at the best of times.

At one time a drought would have sent the pastoralists farther afield, but today the better land, closer to the equator, has become valuable because it is being used to grow crops, and increasingly to grow relatively valuable cash crops. There is nowhere for their animals to feed and such pasture as exists must support far larger numbers of livestock. Instead of being nibbled and allowed to recover and grow again, plants are uprooted and so destroyed. Land is trampled, and when what was never more than marginal semi-desert has lost all its plant life everyone moves on to find the next small patch. Soil from the exposed, parched ground blows away, desert dust and sand joins it on the wind, and the plants on rather better land nearby are buried, and die. The process is repeated endlessly and hopelessly. And so the desert advances.

Ironically, past attempts to improve the lot of the pastoralists have included improvements in the health of their livestock and the development of markets for them. People can earn money by selling animals for slaughter, fewer of their animals die from disease, and that has increased the size of the herds – and of the overgrazing problem.

Hunger and its cure

During the 1930s the volume of world trade was low. There were rich and poor countries, of course, but in the rich countries there were also rich and poor people. The citizens of Europe and America were concerned mainly with their own problems, with the hunger in their own countries rather than hunger in other parts of the world. In any case, before the days of television people received much less information about other countries than they do now. The world problem went largely unnoticed.

When war came vast numbers of educated, sensitive young men and women were conscripted into the armed services and sent overseas. There, especially in Africa and Asia, they saw hunger and poverty on a scale that appalled them. When the war ended those former conscripts vowed that freedom from hunger must be written into a global declaration of human rights. When the United Nations was formed the first of its specialized agencies to come into being, and still the largest, was the Food and Agricultural Organisation (FAO), based in Rome. Its motto is *Fiat Panis* – let there be bread.

At first it must have seemed rather simple. Certain parts of the world produced much more food than they needed, while other parts of the world produced too little. All you had to do was to move food around, and so distribute it more fairly. It would not move of its own accord because the people who needed it could not afford to pay, and the farmers who grew it needed to be paid if they were to stay in business. It would be necessary, therefore, to supply food free, or just for the cost of transporting it, or at any rate for much less than its market value. The United States was the first country to pass legislation to permit such government-subsidized food exports and large amounts were sent to Asia.

It tended to make the problem worse. In the countries receiving food aid large amounts of food became available very cheaply. People bought the cheap food and food grown locally was more difficult to sell because it cost more. It cost more because the farmers who grew it had to be paid its full, unsubsidized price if they were to feed their own families, pay their rents, buy such materials and equipment as they needed, and stay in business. In fact they were often left bankrupt. So local food production fell and the recipient countries became increasingly dependent on the aid.

The Green Revolution

The alternative was to help local farmers to grow more themselves by improving their farming methods. This was the basis of the World Plan for Agricultural Development, drawn up by the FAO. The name was too cumbersome for journalists, who nicknamed the strategy it described the 'green revolution'.

It relied heavily on the introduction of new crop varieties, principally new varieties of wheat bred at the research institute of the Rockefeller Foundation in Mexico and of rice bred at the International Rice Research Institute in the Philippines. The new, so-called 'high-yielding' varieties also received nicknames. They were called 'miracle wheat' and 'miracle rice'. There was nothing miraculous about them, and they formed part of a complex package of improvements.

The Plan had to be based on cereals, for these are the staple foods throughout the world. Traditional varieties of cereal crops grew tall and if fertilizer was added to the soil they grew still taller, producing more straw but not much more grain, and often falling under their own weight so they could not be harvested. The new varieties were short-stemmed and responded to fertilizer applications by producing more stems (tillering) but not longer ones, and much heavier ears, and they ripened more rapidly than the old varieties. Land that once grew one crop a year

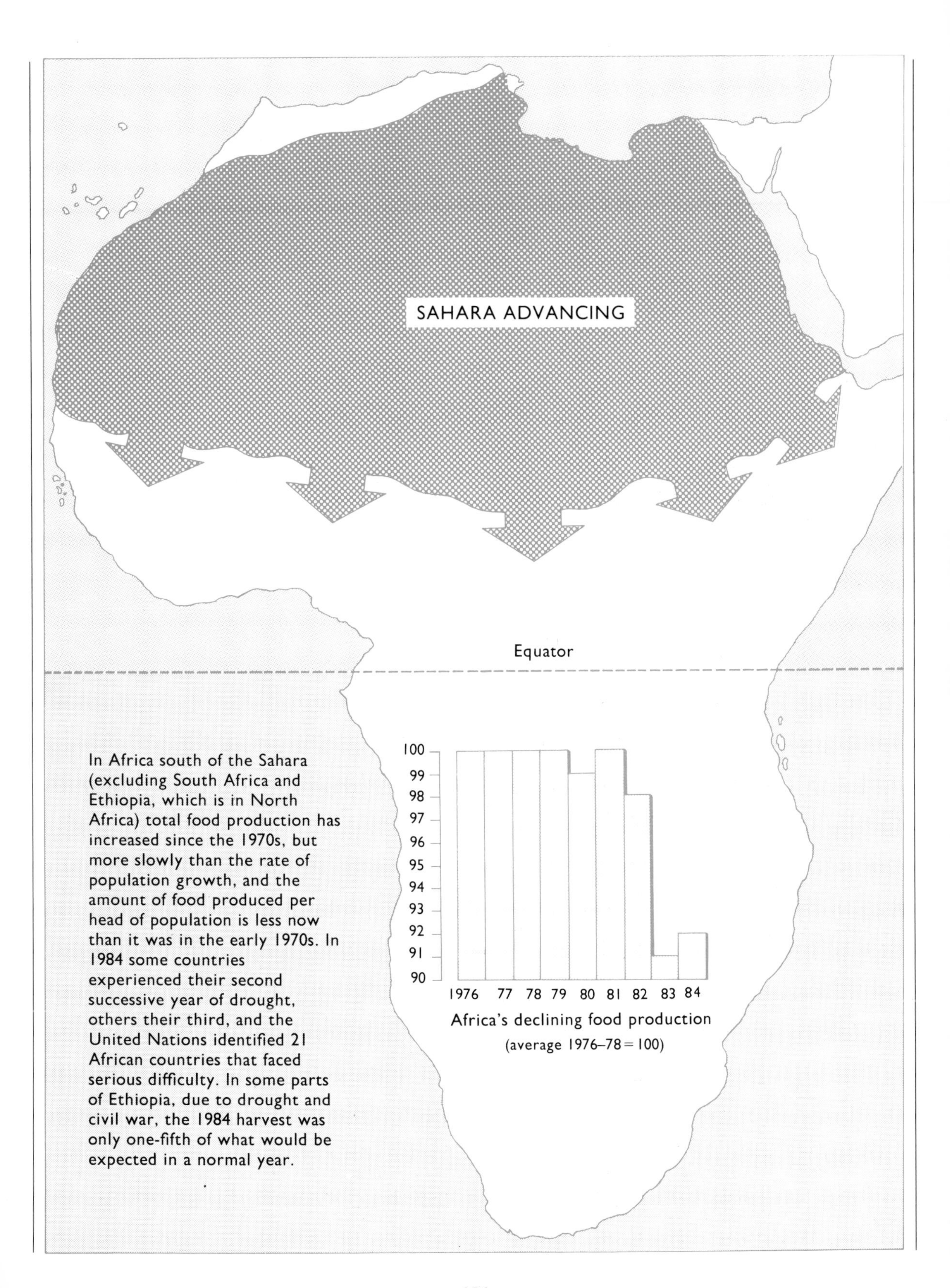

Africa's declining food production
(average 1976–78 = 100)

In Africa south of the Sahara (excluding South Africa and Ethiopia, which is in North Africa) total food production has increased since the 1970s, but more slowly than the rate of population growth, and the amount of food produced per head of population is less now than it was in the early 1970s. In 1984 some countries experienced their second successive year of drought, others their third, and the United Nations identified 21 African countries that faced serious difficulty. In some parts of Ethiopia, due to drought and civil war, the 1984 harvest was only one-fifth of what would be expected in a normal year.

Left *Oasis in Sudan, where water is raised by a method traditional in ancient Egypt.* ***Above*** *In Ethiopia, as in most of Africa, the single-furrow plough is drawn by oxen.* ***Top*** *As cattle and goats eat and trample vegetation around them, settled villages in semi-arid areas become the centres of an expanding wasteland. This is Kordofan province, Sudan.*

could grow two crops and land that formerly grew two crops could now grow three. So far, high-yielding varieties of sorghum and millet – the staple cereals of Africa – have not been developed.

These were the advantages. The disadvantages were formidable. The new varieties were hybrids, and could not be grown from their own seed. Farmers had to buy in seed each year. The new crops needed fertilizer. Without it they were no better than the old varieties, and sometimes they yielded less. The new varieties, grown in denser stands, were more prone to pests and diseases, so they required more skilful care, and that meant pesticides. Larger crops used more water, and in the countries that grew them water was often in short supply.

Farmers who were used to growing enough to feed their own families, with some surplus they traded for things they could not grow or make for themselves, had to move all the way into a cash economy, and most of them needed help. They needed capital to invest in the machines without which they might not be able to manage their larger crops. They needed loans to buy seed, fertilizer and pesticides, which they could repay after the harvest had been sold. They needed better roads and transport, to move bulky materials to and from the villages. They needed wells, pumps, and irrigation equipment.

In some places matters were even worse. Where the farmers were tenants their ability to grow bigger crops made their land more valuable, so landlords increased their rents. Often they raised rents so high that the tenants had to leave, and the land reverted to the landlord, who farmed it himself, hiring his former tenants as labourers on low wages. There was talk of 'the green revolution turning red'. In many countries, though, especially in Asia and to a lesser extent in Latin America, the Plan succeeded. India, which was once desperately and chronically short of food, became a net exporter of food. Other countries were able to improve their situation, or at least to prevent it deteriorating.

Where are the customers?

It was important to improve farming methods, but it did not solve the problem because at its heart there lies the cruellest paradox of all. If the public expresses a demand and if someone is able to supply it, then, provided there is no outside interference with the system, economic theory declares that the supplier and the customers will meet and the demand will be satisfied. If there is insufficient food, therefore, it may be because farmers are unable to meet the demand. Improve their farming methods, allow them to grow bigger crops, and all should be well. If it is not, and large numbers of people still remain hungry, the fault cannot lie on the supply side of the economic equation so it must lie on the demand side.

Farmers cannot afford to give away the food they grow because they, too, have needs. If they are not paid they cannot produce, no matter how many starving people there may be.

If people go hungry while farmers are able to grow as much food as is needed, it can only be because the hungry people are poor and if they are to be fed they will have to find paid work. If they are to find paid work the society to which they belong must start producing goods it can sell in order to provide employment. That is the only way in which an economic demand for food can be stimulated, and that is the paradox. Although many people are hungry, there is too little demand for food and that is why farmers do not grow more.

Structure of populations, by age and sex, for Mexico and Sweden. In Mexico, the large proportion of young people, giving the diagram a pyramid shape, means that population must continue to increase as children grow up and enter the reproductive stage of their lives. In Sweden the more rectangular shape of the diagram indicates that numbers will remain fairly constant.

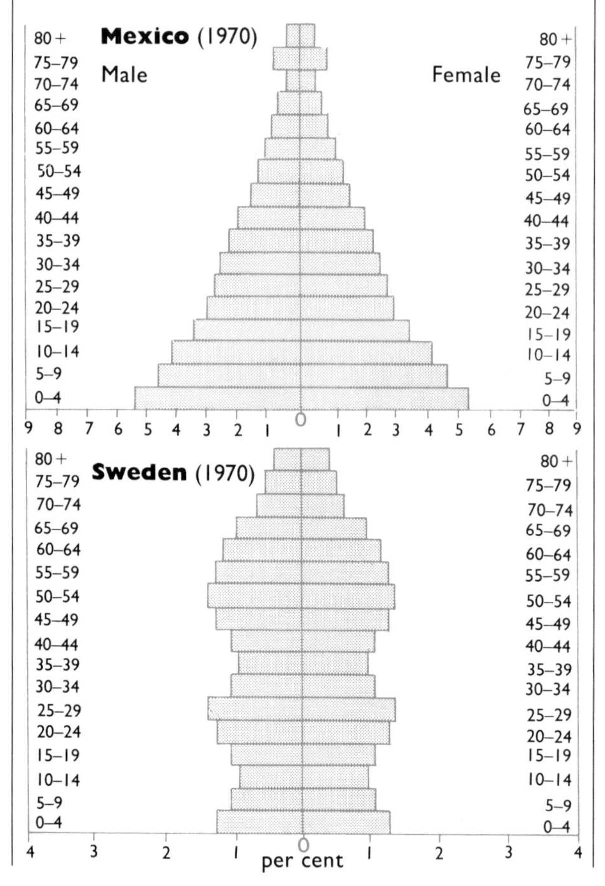

The countries that have succeeded in feeding their people are those that have managed to develop their economies as a whole. It is also in the more successful countries that the rate of population growth is slowing most dramatically.

When people are poor they often rely on their children to support them in their old age, but that may mean having many children because of the high proportion that will die in infancy or childhood. Improved medical services reduce the rate of infant and child mortality, and so for a time populations increase rapidly because there is no reduction in the number of babies born. Then, when people learn that babies have a good chance of surviving to adulthood, they may be willing to have smaller families. At this point couples need to be given a real chance of higher living standards, and educational and employment opportunities for women need to be improved. Where this has happened, as it has in many Asian countries, women have a rewarding alternative to child-rearing, postpone having children for a year or two, and both parents have an incentive to limit the size of the family to one they can support.

The population of the world is still increasing. Not all countries have yet attained this level of development. Earlier high growth rates have produced populations with a high proportion of young people, who will have families of their own. It is estimated that by the end of the century the world population will have increased from its 1983 size of about 4,700 million to about 6,000 million, and that it will stabilize around the middle of the next century at approaching 10,000 million.

What can we do?

Once famine occurs the immediate need is for help, in the form of food, shelter, clothing and medical supplies. Obviously the economic danger of increas-

Famine refugees queueing for food at Korem, northern Ethiopia, in early 1983. Many of these villagers have walked for days to reach Korem. Sustainable improvements must be based on providing better farming methods and paid work for the very poor.

ing dependence through the provision of food does not apply to disaster relief. The task is necessary and deserves every support.

The agencies deal with the supply side of the economic equation, but the demand side also requires urgent attention. We must find ways to make poor people more prosperous, and this is a matter for governments. To the extent that the citizens who elect governments have the power to influence them, that is what we should do. Countries are poor mainly because they must sell primary produce – raw materials – in order to buy secondary produce – manufactured goods – which cost more. In the past the governments of rich countries have tried to help by lending money, but at commercial rates of interest. The loan returned eventually to the lender, with interest. It is not enough. We are all trapped in a kind of game in which everything we do seems to make the rich richer and the poor poorer. How can we break out of the trap?

Probably the most helpful thing would be to import more from the poor countries and pay higher prices for it, while selling our own manufactured goods at less than their full economic price. Economic aid should be given, or at least provided in the form of long-period low-interest or, better still, interest-free loans. This will increase the cost of some of our own raw materials and reduce our visible and invisible export earnings somewhat, but if the poor are to be made more prosperous the rich must expect to suffer some minor inconvenience – and it would be only minor.

What might our help achieve in practical terms? For one thing it might help finance the scheme by which the Sahelian countries plan to reduce overgrazing by developing two belts of grassland stretching most of the way across the continent. Cattle would be raised in the northern drier belt, then taken south, to a region of higher rainfall, to be fattened on better grass. This scheme is part of a wider plan to halt the southward spread of the Sahara by planting trees to bind soil and trap blowing dust and sand. It will work, but it is expensive.

The price of failure

No one can force us to enact reforms. If we fail to do so the disparity between rich and poor will remain and no doubt the suffering we see daily on our television screens will become familiar and we will learn to tolerate it.

The price of our failure, then, will be paid by the poor, but the way they pay it may cause repercussions that do affect us. The spread of the desert is one form of environmental degradation and if it continues can we be sure no further environmental consequences will ensue? Will climates be altered locally? If so, will that alteration extend beyond the immediate desert fringe?

The forests of the humid tropics are being cleared at a rate most scientists agree is alarming. They are being cleared partly to provide land on which poor people may try to grow food for themselves, and partly to provide fuel with which those same people may cook the food they grow. The clearance is caused by poverty, as people seek an escape from hunger.

People will try to obtain food. If the greed of wealthy landowners, or the use of more and more land to grow cash crops crowds food production into a smaller area, then that area will be farmed too intensively, and will deteriorate. Eventually it may be so depleted its recovery takes many years.

If attempts to improve farming methods are not funded adequately, again the land may be damaged. Fertilizers and pesticides used wrongly – through

The green belt proposed by countries bordering the Sahara. It would consist of land on which plants were sown and protected against overgrazing, so helping to check the southward spread of the desert.

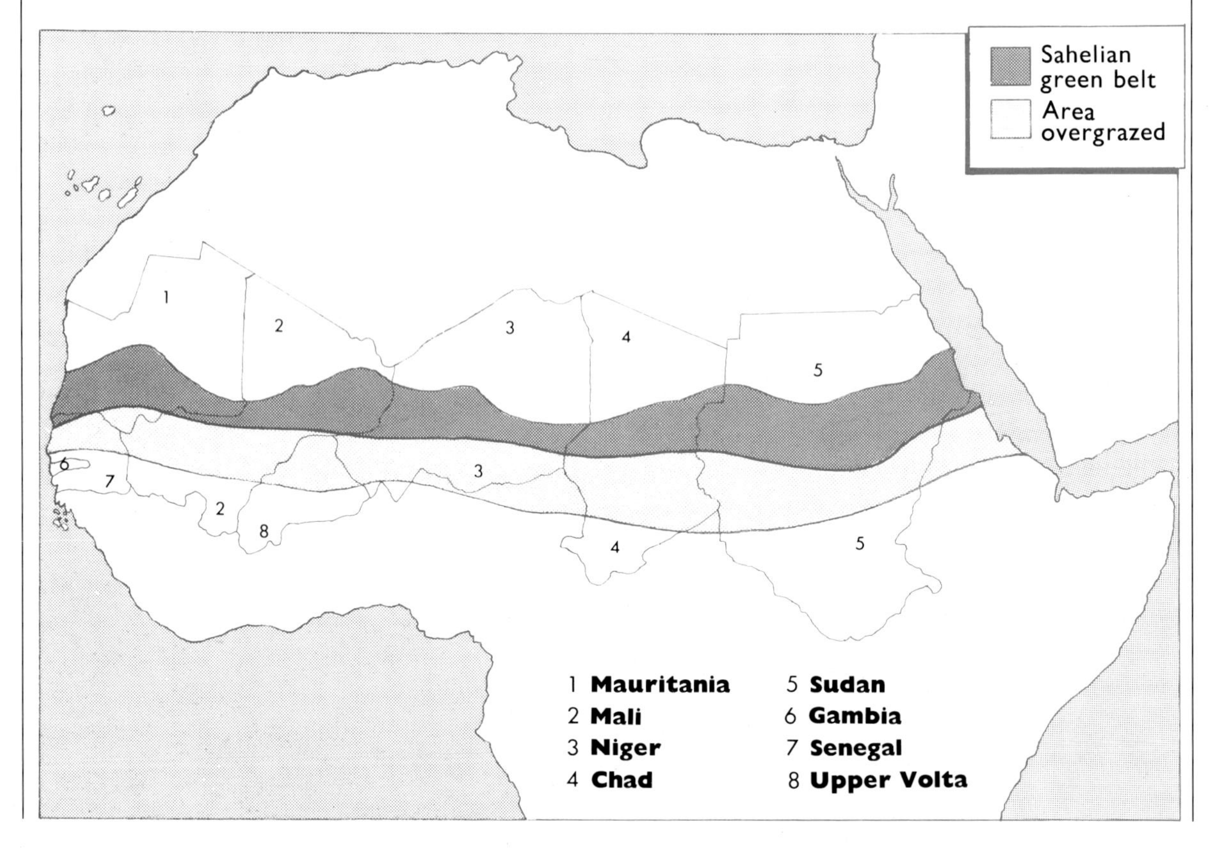

lack of proper equipment or training, for example – can cause serious pollution, especially of water. Irrigation installed without adequate drainage to remove surplus water can lead either to the waterlogging of the land or to its salination, when a high rate of evaporation from the surface leaves behind accumulating mineral salts, dissolved out of the underlying rocks and brought to the surface in solution. Salination sterilizes the land.

Poverty can contribute to the spread of deserts, but it can also create new deserts. It can do so in the humid tropics, in the subtropics, anywhere in fact that people are forced to farm badly or to farm land that should not be farmed.

The problem of hunger and poverty is acute, and it is an environmental problem. Indeed, it may be the most serious of all environmental problems. Yet it remains the problem with which we are most familiar, of which we have greatest experience. We know how to solve it. All we need is the political will to apply our knowledge.

Right *Near Bargan villge, Somalia, rows of brushwood form barriers to check the movement of sand and stabilize dunes.* ***Below*** *Trees also help stabilize dunes, and here prevent them from burying a strip of fertile soil in Majia Valley, Niger.*

Bibliography

Allaby, Michael. 1977. *World Food Resources, Actual and Potential.* Elsevier-Applied Science Publishers, London.

Allaby, Michael and Bunyard, Peter. 1980. *The Politics of Self-Sufficiency.* Oxford University Press, Oxford.

Allaby, Michael and Lovelock, James. 1983. *The Great Extinction.* Secker and Warburg, London.

Bennett, Donald P. and Humphries, David A. 1979. *Introduction to Field Biology.* Edward Arnold, London.

Chauhan, Sumi Krishna. 1983. *Who Puts the Water in the Taps?.* Earthscan-International Institute for Environment and Development, London.

Dajoz, R. 1975. *Introduction to Ecology* (3rd edition). Hodder and Stoughton, London.

Dubos, René. 1980. *The Wooing of Earth.* Athlone Press, London.

Ehrlich, Paul, and Ehrlich, Anne. 1982. *Extinction, The Causes and Consequences of the Disappearance of Species.* Victor Gollancz, London.

Elkington, John. 1980. *The Ecology of Tomorrow's World.* Associated Business Press, London.

Hywel-Davies, Jeremy, and Thom, Valerie. 1984. *The Macmillan Guide to Britain's Nature Reserves.* Macmillan, London.

Independent Commission on International Development. 1980. *North-South: A Programme for Survival* (the report of the Brandt Commission). Pan Books, London.

International Union for Conservation of Nature and Natural Resources (IUCN), with the UN Environmental Programme and the World Wildlife Fund. 1980. *World Conservation Strategy.* IUCN, 1196 Gland, Switzerland.

Lovelock, J.E. 1979. *Gaia, A New Look at Life on Earth.* Oxford University Press, Oxford.

Maitland, Peter S. 1978. *Biology of Fresh Waters.* Blackie, Glasgow.

Myers, Norman. 1979. *The Sinking Ark.* Pergamon Press, Oxford.

Myers, Norman (editor). 1985. *The Gaia Atlas of Planet Management.* Pan Books, London.

Ommanney, F.D. 1971. *Lost Leviathan.* Hutchinson, London.

Owen, Jennifer. 1983. *Garden Life.* Chatto and Windus, London.

Pascarella, Perry. 1979. *Technology, Fire in a Dark World.* Van Nostrand Reinhold, New York.

Pennington, Winifred. 1974. *The History of British Vegetation.* English Universities Press, London.

Rackham, Oliver. 1976. *Trees and Woodland in the British Landscape.* J.M. Dent, London.

Tudge, Colin. 1977. *The Famine Business.* Faber and Faber, London.

Ward, Barbara. 1979. *Progress for a Small Planet.* Maurice Temple Smith, London.

Zisweiler, Vinzenz. 1967. *Extinct and Vanishing Animals.* Longman/Springer Verlag. New York.

Acknowledgments

The Author and Publishers wish to thank the following for their kind permission to reproduce illustrations contained in this book:

Paul Almasy 102; Heather Angel 12, 138b; Biofotos/ Brian Rogers 2–3, 86–7, 90t, 91, 94t, S. Summerhays 170; British Airways 62t; British Antarctic Survey/ J. Parens 78; British Nuclear Fuels 124, 125; British Petroleum Company 164; British Tourist Authority 36; Bruce Coleman 35, 157, Jon & Des Bartlett (Front cover), 137 Mark N. Boulton 6–7, Jane Burton 10–11, Alain Compost 70–71, 90b, 95, A.J. Deane 132, Jessica Ehlers 103, Francisco Erize 83, 97t, M.P.L. Fogden 15, 30, 68, 97b, Jeff Foott 31, 142, 143, Dennis Green 23, Carol Hughes 83b, Gordon Langsbury 18b, Leonard Lee Rue 19, L.C. Marigo 94b, 98, 99, M. Timothy O'Keefe 171, Allan Power 156, Hans Reinhard 104–5, 114, Kim Taylor 18t, Roger Wilmshurst 14, G. Ziesler 32; Central Electricity Generating Board 37; Earthscan/ Mark Edwards 179, 182, 183, 187b, Marcos Santilli 28; English China Clays Group PLC 55; Farmers Weekly/Keith Huggett 9, C. Topham 176; Friends of the Earth 115, David Mansell 62b; Frank Lane Picture Agency/S. Jonasson 66, John Lynch 29, S. McCutcheon 134–5, Mark Newman 139, Robert Pitman (Earthviews) 138t; Geoscience Features Picture Library 82–3, Dr B. Booth 113; Greenpeace 144, 145t, 150, 146, Pereira 151; NASA 59, 73; NHPA/David Woodfall 22; Oxfam 175, 177, Jeremy Hartley 187t; The Photo Source/A.P. 34, 42, 44t, Fox Photos 56–57, Keystone 25, 39, 153, 154–155, T.P. Source 44b, 45; R.S.P.B./ Michael W. Richards 159; Save the Children/ Mike Wells 172–3, 185; Science Photo Library/Martin Bond 131b, Tim Davis 131t, T.P. Dickinson 112, Gazuit (Back cover), Lowell Georgia 106, NASA 67t, National Centre for Atmospheric Research (USA) 65, Robin Scagell 75, Dr Gary Settles 67b, John Walsh 26–7; Seaphot/Planet Earth Pictures: Peter P. Capen, Terra Mar Prod. 166, Colin Doer 167t, Richard Mathews 130, Jesus N. Perez 167b, Warren Williams 163, 162; Swedish Ministry of Agriculture, Swedish Embassy, London 117; Topham Picture Library 40–41, 47, 48, 129, Antman 51; United Kingdom Atomic Energy Authority 118–119, 120, 122, 127.

Index

Figures in italics refer to captions